MAY THE COURSE BE WITH YOU

The Princeton Review is giving away a

FREE

SAT, GMAT, GRE, LSAT, or MCAT course every month.

We'll choose one lucky winner who will be entitled to a free course.

Interested? Simply fill out the card on the right and mail it back to us.

For more information about Princeton Review courses and products, call **(800) REVIEW-6.**

Entries must be received by August 14, 1997. You don't need to buy this book to enter. See official rules on reverse side.

Yes, ...
draw ...

☐ Send m... Princeton Review courses ____________

☐ Also, please send me FREE information about paying for school (including student loan application/information.)

Name: ____________

Address: ____________

____

City: ____________ State: ________ Zip: ________

Is this address ☐ School ☐ Work ☐ Home

Phone: ____________ E-mail address: ____________

School: ____________ Graduation Year: ____________

You should send this card back asap, so it doesn't flap at you every time you open the book.

PLEASE
PLACE
STAMP

THE PRINCETON REVIEW
PO BOX 67
EAST MEADOW NY 11554-0067

OFFICIAL RULES:

We will conduct a random drawing on the fifteenth of every month from all cards we've received between the last drawing and midnight of the day before. We'll hold these drawings through August 15, 1997. The winner of each drawing will receive (at no fee) an SAT, LSAT, GMAT, GRE or MCAT course at any Princeton Review location, each with an approximate value of $695. Your odds of winning depend upon the number of entries received. If you win, you must take your free course within six months of notification; the free course is not transferable except to immediate family members. This promotion is not open to employees of The Princeton Review or Random House and is, of course, void where prohibited by law. All taxes are the sole responsibility of the winners. No purchase is necessary: if the card that's supposed to be attached has already been ripped out, or if you're not buying this book (big mistake), you may enter by sending your own postcard with your name, address, and school to The Princeton Review, 2315 Broadway, New York, NY 10024-4332. You may also write us to get a list of prize winners. By the way, we're not responsible for lost, misdirected, illegible, or mutilated entries.

THE PRINCETON REVIEW

SAT Math workout

Other books in The Princeton Review Series

Cracking the New SAT and PSAT
Cracking the ACT
Cracking the LSAT
Cracking the LSAT with Sample Tests on Computer Disk
Cracking the GRE
Cracking the GRE with Sample Tests on Computer Disk
Cracking the GMAT
Cracking the GMAT with Sample Tests on Computer Disk
Cracking the MCAT
Cracking the SAT II: Biology Subject Test
Cracking the SAT II: Chemistry Subject Test
Cracking the SAT II: English Subject Tests
Cracking the SAT II: French Subject Test
Cracking the SAT II: History Subject Tests
Cracking the SAT II: Math Subject Tests
Cracking the SAT II: Physics Subject Test
Cracking the SAT II: Spanish Subject Test
Cracking the TOEFL with audiocassette
How to Survive Without Your Parents' Money
Grammar Smart
Math Smart
Reading Smart
SAT Verbal Workout
Study Smart
Student Access Guide to America's Top 100 Internships
Student Access Guide to College Admissions
Student Access Guide to the Best Business Schools
Student Access Guide to the Best Law Schools
Student Access Guide to the Best Medical Schools
Student Access Guide to Paying for College
Student Access Guide to the Best 309 Colleges
Student Access Guide to Visiting College Campuses
Trashproof Resumes
Word Smart: Building a Better Vocabulary
Word Smart II: How to Build a More Educated Vocabulary
Writing Smart

Also available on cassette from Living Language

Grammar Smart
Word Smart
Word Smart II

THE PRINCETON REVIEW

SAT Math workout

By Cornelia Cocke

Random House, Inc.
New York 1995

ISBN: 0-679-75363-X

SAT and Scholastic Assessment Test are registered trademarks of the College Entrance Examination Board.

Manufactured in the United States of America on recycled paper

9 8 7 6 5 4 3 2

First Edition

ACKNOWLEDGMENTS

I would like to thank Chris Kensler, Laurice Pearson, Jane Lacher, Jeannie Yoon, Meher Khambata, and Andrea Gordon for their invaluable help, editing skill, and overall perspicacity. For additional production and editing help, thanks to Andrea Paykin, Mike Freedman, Andy Lutz, Lee Elliott, Cynthia Brantley, Julian Ham, Peter Jung, Andrew Dunn, Clayton Harding, Kathleen Standard, Jefferson Nichols, Sara Kane, Ramsey Silberberg, Matthew Clark, Illeny Maaz, Dinica Quesada, Carol Slominski, Christopher J. Thomas, Christopher D. Scott and John Bergdahl.

TABLE OF CONTENTS

INTRODUCTION

READ THIS STUFF FIRST

Sure, the SAT strikes terror in the hearts of test-takers across the land—but it doesn't have to be that way. The SAT supposedly predicts your academic success in college, but we don't think it does anything more than measure one simple thing: your ability to take the SAT.

Your SAT score is not genetic.
Your SAT score is not a measure of your personality or intelligence.
Your SAT score will rise dramatically if you study the right way.

How do we know? Thousands of students taking The Princeton Review course have raised their scores well over 100 points, with some students scoring as much as 250 points higher than their first score. Many of those points come in math, for two reasons: a lot of SAT math questions test basic knowledge, stuff you had a few years ago. A thorough review will prepare you for those questions, so you aren't scratching your head and trying to remember what Mr. Dingdong taught you back in arithmetic class.

The second reason is that The Princeton Review trains you to do only the specific kinds of math questions that actually appear on the SAT, in the quickest and most accurate way possible. If it's not covered in this book, you don't need to know it for the math part of the SAT.

STRUCTURE OF THE MATH SECTIONS

Of the seven sections on the SAT, three of them will be math. (You may have four, in which case one of them will be experimental and won't count towards your score.) There are three different question formats: regular multiple-choice, quantitative comparison, and grid-ins. We'll discuss how to deal with each of those later on. Here's what the format of the three sections looks like:

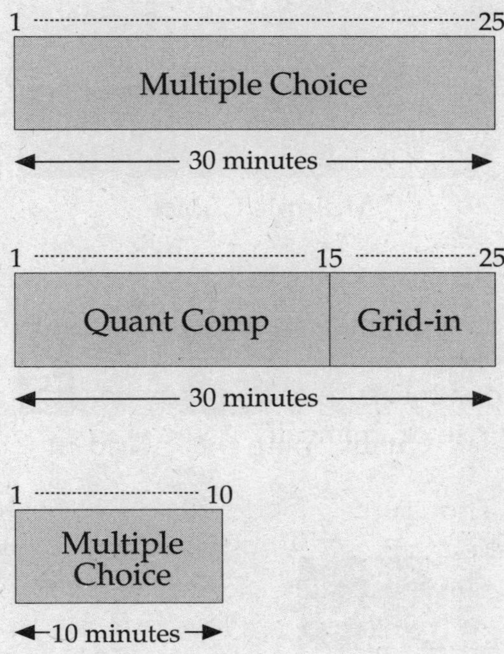

ORDER OF DIFFICULTY

Each section is arranged in order of difficulty, with the easy questions at the beginning, medium questions in the middle, and the hard questions at the end of each section. It's a simple thing, but it gives you a monster advantage in math. This order should guide your guessing *and* your pacing.

When we say "hard," we don't mean that the math involves different and more horrible concepts. We only mean that most test-takers get those questions wrong. And most test-takers get the easy problems right. One of the things you'll notice as you work through this book is that medium and hard questions may simply have more steps in them, so the questions are a little longer—or, especially on hard questions, there's something that steers people to the wrong answer. Don't expect to get a hard question right just by glancing at it. Try to develop a sense of how much work is required to get questions from each category right, and you won't be fooled into picking the trap answers.

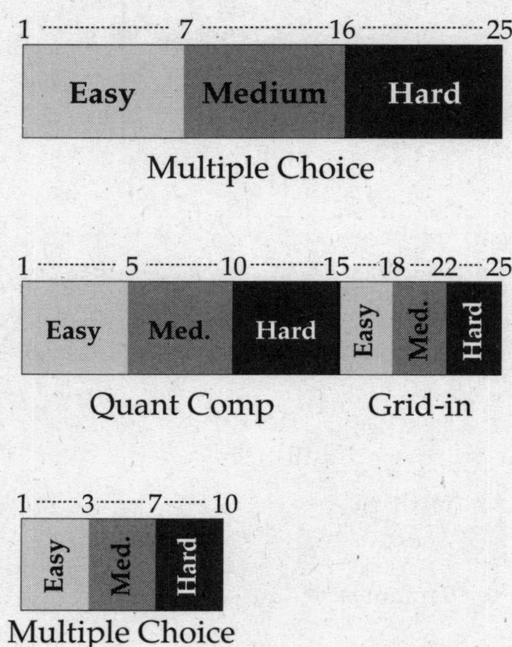

Multiple Choice

Quant Comp Grid-in

Multiple Choice

PACING

Almost everybody works too fast. The SAT isn't your usual math situation—you get *no credit* for partial answers or for understanding the concept of the question. The only thing that matters is what you bubble or grid in on your answer sheet. So take it slow. If you find yourself making careless mistakes, you are throwing points out of the window.

Unless you're shooting for a 700 or above, do not finish the math sections. Again, this isn't like math class. The test isn't designed for you to finish, and you'll hurt your score by trying to do so. If you miss around 5 - 6 questions total for all 3 sections, you're probably hitting the right pace. More mistakes than that, and you're going too fast. If you aren't missing any questions but aren't finishing, you should guess more aggressively and try to work a bit faster.

SCORING

If you were betting your hard-earned cash, wouldn't you want to know the odds? On the SAT, you're betting for more points, and it's important to understand how the scoring works so you'll play smart.

For each right answer, you earn one raw point. For each wrong answer, you lose 1/4 of a raw point. That's it. If you leave a question blank, nothing happens either way, except that the total number of points you can earn is reduced.

Every right answer earns you 1 point, whether it's easy or hard.

That's important to understand, because most people spend too much time on hard questions. They aren't going to do anything more for you than easy questions—and you'll hurt your score if you miss easy or medium questions because you're rushing to finish.

Based on a sample conversion table, here are some examples of what you have to do to get a particular score—there are 60 math questions total. That's a total of 19 easy, 22 medium, and 19 hard questions.

To get a

400: 14 right/0 wrong/46 blank

400: 19 right/12 wrong/29 blank

500: 28 right/0 wrong/32 blank

500: 32 right/8 wrong/20 blank

550: 36 right/0 wrong/24 blank

550: 38 right/8 wrong/14 blank

600: 42 right/0 wrong/18 blank

600: 44 right/8 wrong/8 blank

650: 49 right/0 wrong/11 blank

650: 50 right/ 4 wrong/ 2 blank

700: 54 right/ 0 wrong/ 6 blank

700: 55 right/ 4 wrong/ 1 blank

Amazing, isn't it? Even to get a very high score, *you don't have to finish.* Accuracy is more important than speed! (The examples are based on a sample conversion table; the table for your test may be slightly different—maybe a question or two higher or lower. We put this in just to give you an approximate idea of the score a particular number of right/wrong/blank will give you.)

GUESSING

Say you start to work on a problem and get stuck. Should you just move on? Not if you can cross out any of the answer choices. Are you working on a hard problem? Then cross out any obvious answers and guess from what's left. Are you working on an easy question? Go with your instincts.

If you can eliminate even one answer choice, guess.

Why? Because if you guessed randomly on five questions, without eliminating anything (let your pet monkey pick the answer) you'd have a 1 out of 5 chance of picking a right answer. That one right answer earns you (or your pet monkey) 1 raw point, and the four wrong answers cost you $4 \times \frac{1}{4}$ of a point subtracted. You break even. Eliminating one or more answer choices improves your odds considerably—so take advantage of it!

WHEN TO GUESS

All of that said, we don't mean to suggest that you skip merrily through the sections guessing with abandon. There are smart places to guess, and not-so-smart places to guess.

Always be aware of where you are in the section. Use the order of difficulty information to guide your guessing: easy questions = easy answers; hard questions = hard answers.

GOOD GUESS

- Quantitative Comparison. On hard questions, cross out the obvious answer and guess. (You've got a 1-in-3 shot at it.)
- Geometry, drawn to scale. If you can approximate the length or area or angle measurement, go for it. Applies to easy, medium, and hard questions.
- Any grid-in you've got an answer for. No penalty for wrong answers.

BAD GUESS

- Obvious answers on hard questions.
- Long, complicated word problems at the end of the section (spend your time on something shorter and more manageable).
- Questions you don't have time to read.

CALCULATORS

Seems like a good deal, doesn't it? Well, maybe. It depends on the problem. Don't grab your calculator too quickly—you have to know what to do to solve the problem first. And calculators are only helpful when you have an ugly calculation to do—if the calculation is simple, you might as well do it yourself. Keep in mind that a calculator is no guarantee of anything: you could punch in the wrong number or the wrong operation very easily. It's no crutch, and it won't get you the answer to any question you wouldn't be able to do otherwise.

A calculator will, of course, relieve you of the burden of unpleasant multiplying and dividing.

Think before you punch.

TIPS TO CALCULATOR HAPPINESS

- Get a calculator that follows the order of operations and has keys for x^2, y^x, and $\sqrt{\ }$.
- Use the same calculator every time you practice SAT problems.
- Estimate your answer first.
- Check each number after you punch it in.

CARELESS MISTAKES

If you are prone to careless mistakes, and most of us are, you probably make the same kind of careless mistake over and over. If you take the time to analyze the questions you get wrong, you will discover which kinds are your personal favorites. Then you can compensate for them when you take the SAT.

In the world, and in math class, it's most important for you to understand concepts and ways to solve problems. On the SAT, it's most important that you bubble in the correct answer. Students typically lose anywhere from 30–100 points simply by making careless, preventable mistakes.

Some common mistakes to watch for:

- Misreading the question
- Computation error
- Punching in the wrong thing on the calculator
- On a medium or hard question, stopping after one or two steps, when the question requires three or four steps
- Failing to estimate first
- Answering a different question from the one asked

If, for example, you find you keep missing questions because you multiply wrong, then do every multiplication twice. Do every step on paper, not in your head. If you make a lot of mistakes on positive/negative, write out each step and be extra careful on those questions. Correcting careless mistakes is an easy way to pick up more points—so make sure you analyze your mistakes so you know what to look out for.

CHAPTER 1

Strategies

STRATEGIES

We'll say it again: this isn't the kind of test you get in math class. You need some special techniques for handling SAT problems—techniques that will help you go faster and that take advantage of the format of the questions. Some of the things we suggest may seem awkward at first, so practice them. If you do the math questions on the SAT the way your math teacher taught you, you waste time and throw points away.

1. BACKSOLVING

Good news. Unlike the math tests you usually have in school, the SAT is primarily multiple choice. That means you don't have to come up with an answer on your own—it's going to be one of the five answers sitting right in front of you.

HOW TO RECOGNIZE BACKSOLVING QUESTIONS

- The question will be straightforward—something like "How old is Bob?" "How many potatoes are in the bag?" "What was the original cost of the stereo?"
- The answer choices will be numbers. No variables.
- Your first thought will be to write an equation.

HOW TO SOLVE BACKSOLVING QUESTIONS

Don't write an equation.

Instead, pick an answer and work it through the steps of the problem, one at a time, and see if it works.

In essence, you're saying *what if C is the answer, does that solve the problem?*

Here's an example:

If $\frac{3(x-1)}{2} = \frac{9}{x-2}$, what is the value of x?

(A) –4 (B) –2 (C) 1 (D) 4 (E) 9

Solution: Let's try C first. We'll plug in 1 for x and see if the equation works:

$$\frac{3(1-1)}{2}=\frac{9}{1-2}$$
$$\frac{0}{2}=\frac{9}{-1}$$

OK, so C isn't the answer. Cross it out. Let's try D:

$$\frac{3(4-1)}{2}=\frac{9}{4-2}$$
$$\frac{9}{2}=\frac{9}{2}$$

The equation works, so D is the answer. Sure, we could have done the algebra, but wasn't backsolving easier? What we've done is turn an algebra problem into an arithmetic problem, and all we have to do is manage simple operations like 4 – 1. You're much more likely to make mistakes dealing with x than with 4 – 1. Also, when you backsolve, you're taking advantage of the fact that there are only 5 answer choices. One of them is correct. You might as well try them and find out which one it is—and you no longer have to face the horror of working out a problem algebraically and finding your answer isn't one of the choices.

Here's a harder example:

> Pinky had twice as many potatoes as Dan, who had the same number of potatoes as Zippy. If Pinky gives 5 potatoes to Zippy, then Dan has three times as many potatoes as Pinky. How many potatoes did Dan have?
>
> (A) 10 (B) 6 (C) 3 (D) 2 (E) 1

Solution: Let's try C first. If Dan started with 3, then Zippy also started with 3. Since Pinky had twice as many potatoes as Dan, then Pinky started with 6. If he gives 5 to Zippy, he now has 1 and Zippy has 8. Pinky has 1 potato and Dan has 3 potatoes, or three times as many as Pinky. So C is the right answer.

To keep things organized, make a chart:

	D	Z	P
Originally:	3	3	6
After exchange:	3	8	1

TIPS FOR BACKSOLVING HAPPINESS

- C is a good answer to try first, unless it's awkward to work with.
- The answers will be in numerical order, and you may be able to eliminate answers that are either too big or too small, based on the result you got with C.
- Don't try to work out all the steps in advance—the nice thing about backsolving is that you do the steps one at a time.
- Backsolving questions may be long word problems or short arithmetic problems, and they can appear in the easy, medium, or difficult sections. The harder the question, the better off you'll be backsolving it.

Make a chart if you have a lot of stuff to keep track of.

On the next page is a Quick Quiz, so you can practice backsolving before you continue. The question number corresponds to the difficulty level in the 25-question multiple-choice section (see Order of Difficulty, page 3). Answers and explanations immediately follow every Quick Quiz.

QUICK QUIZ #1

EASY

6 If 4 less than the product of b and 6 is 44, what is the value of b?

(A) 2
(B) 4
(C) 6
(D) 8
(E) 14

MEDIUM

13 A store reduces the price of a CD player by 20%, and then reduces that price by 15%. If the final price of the CD player is $170, what was its original price?

(A) $140
(B) $185
(C) $200
(D) $250
(E) $275

HARD

24 Triangle ABC has sides measuring 2, 3, and r. Which of the following is a possible value for r?

(A) 0.5
(B) 1
(C) 2
(D) 5
(E) 6

Answers and Explanations: Quick Quiz #1

6 **D** Let's try C first, so $b = 6$. The product of 6 and 6 is 36, and 4 less than 36 is 32. 32 isn't 44—cross out C. Let's try a higher number, like D. If $b = 8$, the product of 8 and 6 is 48, and 4 less than 48 is 44. Yeah!

13 **D** Try C first. If the original price of the CD player was $200, then 20% of 200 is 40. That leaves us with a price of $160. Hey—the final price was $170, and we're already below that. We need a higher number. Try D: if the original price was $250, take 20% of 250 = 50. Now the price is $200. Take another 15% off and we get 250 – 30 = 170.

32 **C** You need to know a rule here—the sum of any 2 sides of a triangle must equal more than the third side. Try C first. If $r = 2$, then the sides are 2, 2, 3. Add up any pair and you get a number that's higher than the remaining number. So it works. (Try some of the other answers, just for practice, and see how they *don't* work.)

Let's do a little analysis. See how the questions got harder as you went along? In the easy question, you had to read carefully, multiply, and subtract. In the medium question, you had to take percentages. In the hard question, you had to deal with geometry without a diagram, and also know a particular rule. For all the questions, backsolving allowed you to avoid writing an equation. Less work is good.

---

2. PLUGGING IN

This technique is exactly like backsolving, but you use it for algebra questions—usually ones that have variables in the answer choices. That means you have to make up your own number to substitute into the problem, rather than use the answer choices.

HOW TO RECOGNIZE A PLUGGING-IN QUESTION

- Variables in the answer choices
- The question says something like *in terms of x*...
- Your first thought is to write, or solve, an equation.

- The question asks for a percentage or fractional part of something, but doesn't give you any actual amounts.

HOW TO SOLVE A PLUGGING-IN QUESTION

- Don't write an equation.
- Pick an easy number and substitute it for the variable.
- Work the problem through and get an answer. Circle it so you don't lose track of it.
- Plug your number—the one you chose in the beginning—into the answer choices and see which choice produces your circled answer.

Here's an example:

Pinky spent x dollars on pet toys and 12 dollars on socks. If the amount Pinky spent was twice the amount she earns each week, how much, in terms of x, does Pinky earn each week?

(A) $2(x + 12)$

(B) $2x + 24$

(C) $\frac{x}{2} + 12$

(D) $\frac{x+12}{2}$

(E) $\frac{x-12}{2}$

Solution: Let's plug in 100 for x. That means Pinky spent a total of 112 dollars. If that was twice her weekly salary, then she makes half of 112, or 56 dollars a week. Circle *56*. Now we plug our number, 100, into the answers to see which one gives us 56. (A) $2(100 + 12) = 224$. No good. (B) too big. (C) $50 + 12 = 62$ (D) $\frac{112}{2} = 56$! Yes! (E) way too big. The answer is D.

Here's a harder example:

Ziggy bought x cans of red goo for y dollars a can, and z cans of blue goo for $3y$ dollars a can. If he bought twice as much blue goo as red goo, then in terms of y, what was the average cost, in dollars, per can of goo?

(A) $\frac{3y}{2}$ (B) $\frac{7y}{3}$ (C) $3y - y$ (D) $2y$ (E) $6y$

Solution: You don't really want to do the algebra, do you? Let's use simple, low numbers and plug in. How about $x = 2$ and $y = 3$? That's 2 cans of red goo at \$3 each. So he spent \$6 on red goo. (In word problems, it helps to keep track of what the numbers represent.) If he bought twice as much blue goo, then $z = 4$. So he bought 4 cans of blue goo at $3y$ or \$9 a can, and spent a total of \$36. Now we figure the average price by adding up the dollars spent and dividing that by the total number of cans. He spent \$6 + \$36 = \$42 on 2 + 4 = 6 cans of goo. So the average price per can is $\frac{42}{6}$ = \$7. Circle \$7.

Now, since we said $y = 3$, plug 3 in for y in the answer choices and see which one gives you 6. Choice A is 9. B is 7, so that's the answer.

Here's a different kind of example:

Mr. Heftwhistle gave 30% of his socks to a sockless friend. Mr. Heftwhistle then decided to move to Florida where he would rarely need socks, so he gave 60% of the socks that remained to another friend who had a sock-loving pet. What percent of his original number of socks did Mr. Heftwhistle have left?

(A) 10%
(B) 18%
(C) 28%
(D) 36%
(E) 40%

Solution: If you don't plug in, you may make the sad mistake of picking A. We wouldn't want that. Let's say Mr. Heftwhistle had 100 socks. 30% of 100 = 30, so he's left with 70. 60% of 70 = 42, so he's left with 28. Here's the great thing about plugging in 100 on percentage problems—28 (left) out of 100 (original number) is simply and happily and easily 28%. That's it. C is the answer.

TIPS FOR PLUGGING-IN HAPPINESS

- Pick easy numbers like 2, 4, 10, 100. The best number to choose depends on the question—100 for percents, 12 for feet/inches, etc.
- Avoid picking 0, 1, or any number that shows up in the answer choices.
- If the number you picked leads to ugly computations—fractions, negatives, or anything you need a calculator for—bail out and pick a different number.
- Practice!

QUICK QUIZ #2

EASY

4 If p is odd, which of the following must also be odd?

(A) $p + 1$

(B) $\frac{p}{2}$

(C) $p + 2$

(D) $2p$

(E) $p - 1$

MEDIUM

12 If $\frac{y}{3} = 6x$, then in terms of y, $x =$

(A) $3y$ (B) $2y$ (C) y (D) $\frac{y}{2}$ (E) $\frac{y}{18}$

HARD

19 Mary spilled $\frac{2}{5}$ of her peanuts, and Dolly ate $\frac{1}{3}$ of what was left. Dolly then gave the peanuts to Max and Bob, who each ate half of what remained. What fractional part of Mary's peanuts did Bob eat?

(A) $\frac{1}{15}$ (B) $\frac{1}{10}$ (C) $\frac{1}{5}$ (D) $\frac{1}{3}$ (E) $\frac{4}{5}$

Answers and Explanations: Quick Quiz #2

4 C p has to be odd. Let's make $p = 3$. Try that in the answer choices, and cross out anything that isn't odd. Choice A: $3 + 1 + 4$. Cross out A. B: $\frac{3}{2}$ Cross out B. (Fractions can't be odd or even.) C: $3 + 2 = 5$, leave C in. D: $2(3) = 6$. Cross out D. E: $3 - 1 = 2$. Cross out E.

12 E Plug in $y = 36$, which makes $x = 2$. Now we plug in 36 for y in the answer choices and look for x, which is 2. Choice A: something huge. B: still something huge. C: 36. D: $\frac{36}{2} = 18$, still too big. E: $\frac{36}{18} = 2$, so E is correct.

19 C On this kind of question, there aren't variables in the answer choices, but there's an *implied* variable in the question, because we don't know how many peanuts Mary started with. Let's say she had 10 ounces of peanuts. (The "ounces" doesn't really matter—it's the number that's important.) If she spilled $\frac{2}{5}$ of 10, she spilled 4, leaving her with 6. If Dolly ate $\frac{1}{3}$ of 6, she ate 2, leaving Mary with 4. If Max and Bob split 4, they each ate 2. The fractional part is $\frac{\text{part}}{\text{whole}}$, so Bob's fractional part is $\frac{2}{10}$ or $\frac{1}{5}$. (Whew.)

A tree diagram makes this easier to deal with:

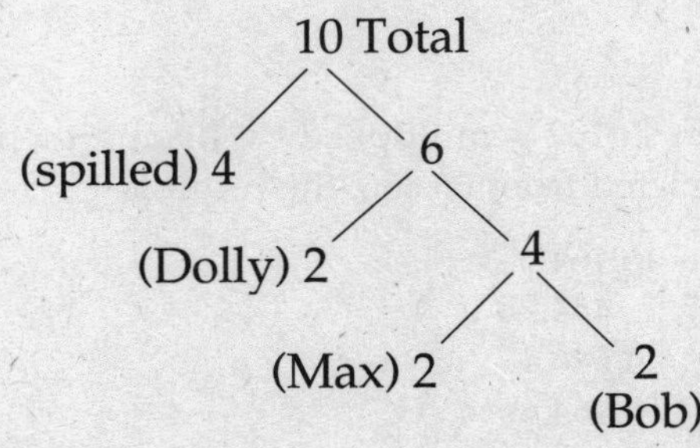

TIP: In Question 12, you may have had a hard time coming up with numbers that worked evenly. That's OK—it takes practice. You can plug in any numbers you want, as long as they satisfy the conditions of the problem, so you might as well plug in numbers that are easy to work with.

TIP: In Question 19, you could solve this without plugging in, but then you're dealing with fractional parts of a whole, the whole being Mary's peanuts. It's very easy to get confused doing it that way, because the numbers aren't concrete and they quickly become meaningless. The advantage of plugging in is that you're working with actual amounts, just like real life. One more thing—we picked 10 because the first thing we had to do was take $\frac{2}{5}$ of it. If we'd picked a number we couldn't take $\frac{2}{5}$ of easily, we would have tried another number.

---

3. ESTIMATING

ON ARITHMETIC QUESTIONS, GETTING A ROUGH ESTIMATE OF THE ANSWER MAY BE ALL THAT'S NECESSARY.

We like that. The less work the better. Maybe you'll only be able to eliminate a couple of answers. That's OK too.

For example:

When .20202 is multiplied by 10^5 and then subtracted from 66,666, the result is

(A) –46,464
(B) 464.98
(C) 4,646.4
(D) 6,464.6
(E) 46,464

Solution: First multiply .20202 by 10^5. Just move the decimal point 5 places to the right. You get 20,202. (Use your calculator if you want.) Now you're going to subtract that from 66,666—but estimate it before you continue. Looks to be around 40,000 or so, doesn't it? So pick E and go on.

Two advantages: one, you avoided having to do the last step of the problem and gained yourself some time, and two, you avoided even the possibility of making a careless mistake in that last step.

We know you can subtract. That's not the issue. On a timed test, with a lot of pressure on you, the fewer steps you have to do, the better off you are.

GEOMETRY FIGURES ARE DRAWN TO SCALE UNLESS SPECIFICALLY STATED OTHERWISE.

This is a fabulous piece of news—it means that you should use your eyes to estimate distances and angles, instead of jumping immediately to formulas and equations. You aren't allowed to bring a ruler or a protractor into the test. But you can often tell if one line is longer than another, or if the shaded part of a circle is larger than the unshaded part, just by estimating. That should allow you to eliminate at least a couple of answers, maybe more.

Is this a sleazy technique? Are we telling you to take the easy way out? No and yes. ETS, the company that writes the SAT, doesn't mind if you use your common sense. Neither do we. And as for the easy way out...yes, that's exactly what you're training yourself to look for.

Estimating is not totally foreign to you. Think of geometry problems you encounter in real life—parking a car, packing a box, even shooting a basketball. We guess you don't take out a pad and pencil and start calculating to solve any of these problems. You estimate them, and see what happens. Same deal on the SAT.

For example:

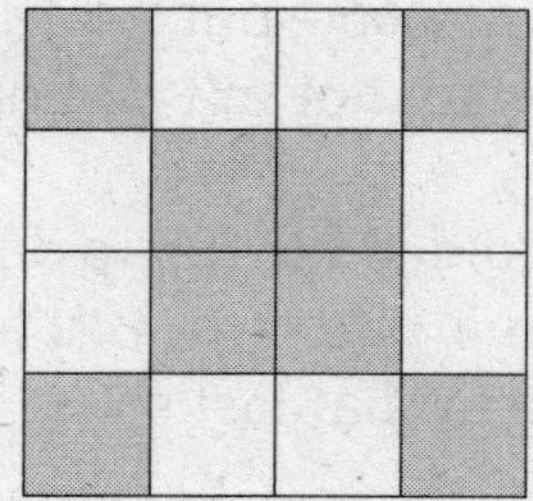

What fractional part of the square is shaded?

(A) $\frac{1}{4}$

(B) $\frac{3}{10}$

(C) $\frac{1}{2}$

(D) $\frac{7}{12}$

(E) $\frac{15}{16}$

8/16 = 2/4 = 1/2

Solution: Just look at it. How much looks shaded? A little? No, so cross out A and B. Most of it? No, so cross out E. That leaves you with two answer choices, which isn't bad, since you haven't done any math. If you get stuck here, guess. Or count up how many shaded squares there are, and put that over the total number of squares. So the fractional part is $\frac{8}{16}$, or $\frac{1}{2}$. The answer is C.

QUICK QUIZ #3

EASY

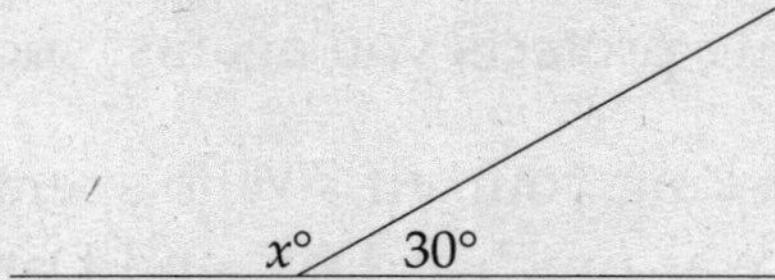

4 Which of the following is equal to $3x$?

(A) 50 (B) 120 (C) 150 (D) 360 (E) 450

MEDIUM

13 Dan, Laura and Jane went grocery shopping. Dan spent three times as much as Laura and half as much as Jane. If they spent a total of $50 on groceries, how much did Jane spend?

(A) $15 (B) $20 (C) $25 (D) $30 (E) $45

HARD

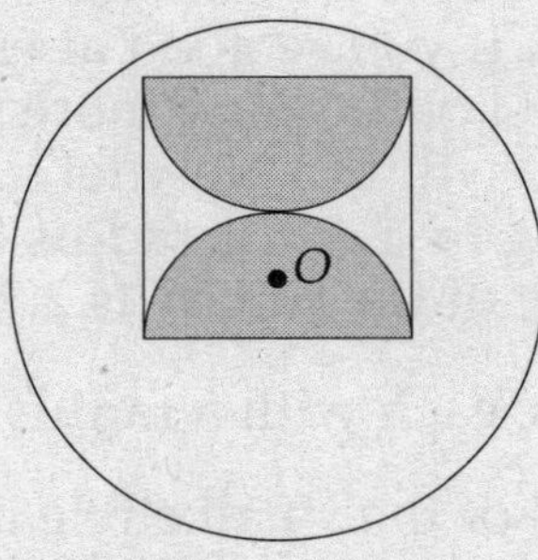

17 In the figure above, the radius of the circle is equal to the length of a side of the square. If the shaded region represents two semicircles inscribed in the square, the ratio of the area of the shaded region to the area of the circle is

(A) 1:16 (B) 1:8 (C) 1:4 (D) 1:2 (E) 2:3

Answers and Explanations: Quick Quiz #3

4 E Angle x is pretty big, isn't it? So $3x$ is really really big. Cross out A, B, and C. Now work it out. $x + 30 = 180$, so $x = 150$. And $3x = 450$. If you fail to estimate, you might forget to multiply by 3, and pick C. You might fall asleep for a split second and divide by 3, and pick A. Estimating protects you against such disasters.

13 D You might start by asking yourself, "Who spent the most money?" since Jane spent twice as much as Dan, and Dan spent three times as much as Laura, Jane spent the most. You can definitely eliminate A; it's too small an amount for Jane to have spent. Now backsolve. Begin with C or D since A is out. Which is the easier number to cut in half?

	J	D	L
(D)	30	15	5

\$30 + \$15 + \$5 = \$50, so D is your answer.

17 C Look at the figure. How much of it looks shaded? Less than half? Sure. Cross out D and E. If you're good at estimating, maybe you can cross out A as well. (Try drawing more semi-circles in the big circle and see how many will fit.) Now let's figure it out, using our good friend plugging in: Let the radius = 2. So the area is 4π. If the radius = 2, the side of the square is 2. The shaded part consists of 2 semi-circles, each with a radius that's $\frac{1}{2}$ the side of the square, so the radius of the small circle is 1, and the area is π. Put the small area over the big area and you get $\frac{\pi}{4\pi} = \frac{1}{4}$, which is a ratio of 1:4.

TIPS FOR ESTIMATING HAPPINESS

- With geometry, especially on hard questions, the answer choices need to be translated into numbers that you can work with.
- Translate π to a bit more than 3; $\sqrt{2}$ is 1.4; $\sqrt{3}$ is 1.7.
- Practice estimating *a lot*, even if you're going to work the problem out—and notice how your estimates improve.
- The farther apart the answer choices, the bigger the opportunity for eliminating answers by estimating.
- If the figure is NOT drawn to scale, redraw it if you can, using whatever measurements are provided. Then go ahead and estimate. If you can't redraw it, don't estimate.
- If two things look about equal, you can't assume that they're *exactly* equal.
- Trust what your eyes tell you.

HOW TO APPLY THESE TECHNIQUES

To study efficiently for the SAT, you must

- Learn how to plug in, backsolve, and estimate in your sleep.
- Be able to recognize plugging in and backsolving questions when they appear.
- Analyze your work so that you can avoid making the same mistakes over and over.
- Do the problems in this book as though you are taking the real thing—practice with the same focus and intensity you will need on the actual SAT.

CHAPTER 2

Arithmetic

ARITHMETIC

There are only three categories—arithmetic, algebra, and geometry—and most of this stuff you learned when you were still in diapers. So most of your work is going to be review. But like everybody else, Michael Jordan and Michael Stipe included, you need to practice. After each section of review is a Quick Quiz: three questions, one easy, one medium, one hard. If you get all three right, you're in very good shape. And keep in mind: you could leave *all* the hard questions blank and still get around a 590. So make sure you're getting the easy and medium ones right first.

More SAT problems involve arithmetic than any other type of math, and you've been doing it since grade school. Practice. You want doing arithmetic to be second nature—it will make working on the hard questions much, much easier. Plus, you'll be able to balance your checkbook, leave the correct tip, and be a better shopper.

1. DEFINITIONS

sum what you get when you add two numbers together

difference what you get when you subtract one number from another

quotient what you get after dividing one number into another

product what you get when you multiply two numbers together

remainder what's left over if a division problem doesn't work out evenly

prime a number divisible evenly only by itself and 1 (example: 2, 3, 5, 7...)

odd a number not evenly divisible by 2 (1 is *not* prime)

even a number evenly divisible by 2 (0 is even)

integer any number that isn't a fraction, including 0

multiple a bigger number that your number goes into (example: 8 is a multiple of 2)

factor same meaning as "division": a smaller number that goes into your number (example: 2 is a factor of 8)

distinct different (i.e., the distinct factors of 4 are 1, 4, and 2; not 1, 4, 2, and 2)

consecutive numbers in order (1, 2, 3 etc.)

numerator the top number of a fraction

denominator the bottom number of a fraction

reciprocal whatever you multiply a number by to get 1. (i.e., the reciprocal of $\frac{1}{2}$ is $\frac{2}{1}$. The reciprocal of 6 is $\frac{1}{6}$.)

PEMDAS you won't see that written on the test—it's a handy acronym for the order of operations: Parentheses, Exponents, Multiplication, Division, Addition, Subtraction. Learn it, live it.

digit a number from 0 to 9. For example, 376 is a three-digit number.

places in 234.167, 2 is the hundreds' place, 3 is the tens' place, 4 is the ones' or units' digit, 1 is the tenths' place, 6 is the hundredths' place, and 7 is the thousandths' place.

TIP: Imaginary numbers and natural numbers are not on the SAT.

QUICK QUIZ #4

EASY

1 What is the greatest common prime factor of 32 and 28?

(A) 1
(B) 2
(C) 3
(D) 4
(E) 7

MEDIUM

9 If x is a positive integer greater than 1, and $x(x + 4)$ is odd, then x must be

(A) even
(B) odd
(C) prime
(D) a factor of 8
(E) divisible by 8

HARD

24 The units' digit of 2^{33} is how much less than the hundredths' digit of $\frac{567}{1000}$?

(A) 1
(B) 2
(C) 3
(D) 4
(E) 5

Answers and Explanations: Quick Quiz #4

1 **B** Backsolve—why bother thinking up the answer yourself when they give you 5 choices? Since the question asks for the greatest common prime factor, you can cross out anything that isn't prime—get rid of A and D. Now start with E, because it's the greatest answer choice. Does 7 go into 32? No. Cross out E. C: Does 3 go into 28? No, cross out C. That leaves us with B. Does 2 go into 32 and 28? Yep.

9 **B** Plug in. If x has to be a positive integer greater than 1, try $x = 2$. But $2(2 + 4)$ isn't odd. So try $x = 3$. $3(3 + 4) = 21$, so that works. Now you can cross out everything but B and C. Try an odd number that isn't prime, say $x = 9$. $9(9 + 4) = 117$, which is odd. So cross out C.

24 **D** The units' digit of any number has a pattern as the exponent goes up. $2^1 = 2$. $2^2 = 4$. $2^3 = 8$. $2^4 = 16$. $2^5 = 32$. $2^6 = 64$. See the pattern? The units' digit of 2^x goes in a pattern of 2, 4, 8, 16 and then repeats forever, all the way to 2^{33} and beyond. So divide the number of elements in the pattern, which is 4, into 33. That gives you 8 with a remainder of 1. That means the pattern fully repeats 8 times, and then starts the pattern again with the first number. So the units' digit of 2^{33} is 2.

To figure out the hundredths' digit of $\frac{567}{1000}$, divide it on your calculator. You should get 0.567, which makes 6 the hundredths' digit. 2 is 4 less than 6, so the answer is D. (When we call these hard questions, we aren't kidding.)

2. DIVISIBILITY

Shortcuts are always good, right? If a question asks about divisibility, use your calculator.

TIP: You can't divide any number by 0.

Factoring shows up on the SAT all over the place. That's OK, it's easy.

To find the prime factors of a number, make a tree, starting with any two numbers that multiply together to equal your original number:

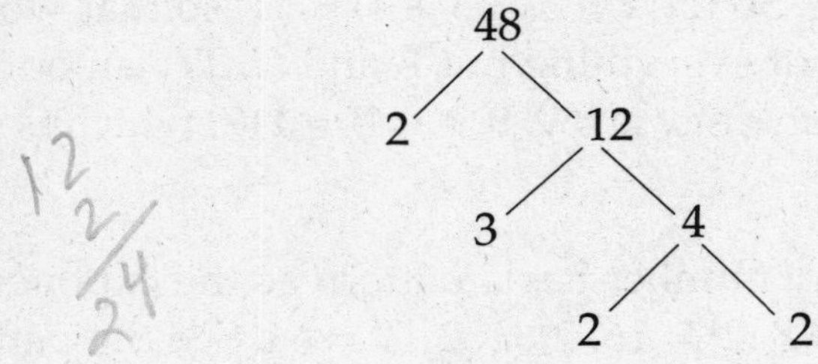

All of the numbers in the tree (2, 12, 3, 4) are *factors* of your original number. The final row of numbers (2, 3, 2, 2) are the *prime factors*.

To find all of the factors of a number, factor in pairs. Start with 1 and make a list of all the pairs that multiply together to equal your original number:

What are the factors of 36?

1, 36
2, 18
3, 12
4, 9
6, 6

QUICK QUIZ #5

EASY

2 Which of the following could be a factor of $n(n+1)$, if n is a positive integer less than 3?

(A) –1
(B) 3
(C) 5
(D) 8
(E) 9

MEDIUM

10 If Darlene divided 210 chocolate kisses into bags containing the same number of kisses, each of the following could be the number of kisses per bag EXCEPT:

(A) 35
(B) 21
(C) 20
(D) 15
(E) 14

HARD

19 If $p = 2^2 \bullet 3^2 \bullet 7$, and y is a positive integer, what is the greatest number of values for y such that $\frac{p}{18y}$ is an integer?

(A) 2
(B) 3
(C) 4
(D) 5
(E) 6

Answers and Explanations: Quick Quiz #5

2 **B** Plug in. If $n = 1$, $1(1 + 1) = 2$. None of the answers are factors of 2. If $n = 2$, $2(2+1) = 6$. 3 is a factor of 6, so the answer is B.

10 **C** You could use your calculator and divide each answer into 210, and pick the one that doesn't go evenly. Or you could factor 210 as $7\times5\times3\times2$. Now factor the answers. A: 7×5, which goes in, so cross it out. B: 7×3. Cross it out. C: $2\times2\times5$. That doesn't go in because there's only 1 factor of 2 in 210.

19 **C** Let's write this a slightly different way: $\frac{2\times2\times3\times3\times7}{2\times3\times3y}$. Cancel, and you get $\frac{14}{y}$. How many different numbers can y be, if $\frac{14}{y}$ is an integer? $y = 1, 2, 7,$ or 14. Four values.

3. FRACTIONS

To add fractions, get a common denominator and then add across the top:

$$\frac{1}{2}+\frac{2}{3}=\frac{3}{6}+\frac{4}{6}=\frac{7}{6}$$

To subtract fractions, it's the same deal, but subtract across the top:

$$\frac{3}{4}-\frac{1}{3}=\frac{9}{12}-\frac{4}{12}=\frac{5}{12}$$

To multiply fractions, cancel if you can, then multiply across, top and bottom:

$$\frac{1}{2}\bullet\frac{3}{5}=\frac{3}{10} \qquad \frac{2}{\cancel{7}_{1}}\bullet\frac{\cancel{14}^{2}}{19}=\frac{4}{19}$$

To divide fractions, flip the second one, then multiply across, top and bottom:

$$\frac{2}{3}\div\frac{1}{2}=\frac{2}{3}\bullet\frac{2}{1}=\frac{4}{3}$$

To see which of two fractions is bigger, cross-multiply from bottom to top. The side with the bigger product is the bigger fraction.

$$55 \quad \frac{5}{7} \times \frac{8}{11} \quad 56$$

56 is bigger than 55, so $\frac{8}{11}$ is bigger.

QUICK QUIZ #6

EASY

3 Which of the following is greatest?

(A) $\frac{3}{5} \times \frac{5}{3} =$

(B) $\frac{3}{5} \div \frac{5}{3} =$

(C) $\frac{3}{5} + \frac{3}{5} =$

(D) $\frac{5}{3} - \frac{3}{5} =$

(E) $\frac{5}{3} \div \frac{3}{5} =$

MEDIUM

13 Boris ate $\frac{1}{2}$ a pizza on Monday and $\frac{2}{3}$ of the remainder on Tuesday. What fractional part of the pizza was left?

(A) $\frac{3}{5}$ (B) $\frac{1}{3}$ (C) $\frac{1}{4}$ (D) $\frac{1}{6}$ (E) $\frac{1}{12}$

HARD

20 At a track meet, $\frac{2}{5}$ of the first-place finishers attended Dingdong High School, and $\frac{1}{2}$ of them were girls. If $\frac{2}{9}$ of the first-place finishers who did NOT attend Dingdong High School were girls, what fractional part of the total number of first-place finishers were boys?

(A) $\frac{1}{9}$ (B) $\frac{2}{15}$ (C) $\frac{7}{18}$ (D) $\frac{3}{5}$ (E) $\frac{2}{3}$

Answers and Explanations: Quick Quiz #6

3 **E** You can do some estimating here, but be careful—remember that when you divide by a fraction, you're really multiplying by the reciprocal. E is $\frac{5}{3} \div \frac{3}{5}$, which is $\frac{5}{3} \times \frac{5}{3} = \frac{25}{9}$. A: 1. B: $\frac{9}{25}$. C: $\frac{6}{5}$. D: $\frac{16}{25}$. If you had trouble figuring any of those out, go back to the fraction review.

13 **B** Let's plug in, and say that Boris's pizza had 6 slices. If he ate $\frac{1}{2}$, then he ate 3 slices and had 3 slices left. If he then ate $\frac{2}{3}$ of 6, he ate 4 slices and had 2 slices left. So the final fractional part is 2, over the original whole of 6, or $\frac{2}{6} = \frac{1}{3}$. How did we know to plug in 6? We looked at the denominators of the fractions we were going to have to work with, and picked a number that they would go into evenly. If you pick a bad number, don't worry, just pick a new one.

20 **E** Very complicated question. Let's plug in. If the total number of first-place finishers was 30, then $\frac{2}{5}$ of 30 = 12 (from Dingdong High School). That leaves 18 who did not go to Dingdong High School. If half the Dingdong runners were girls, that means 6 were girls and 6 were boys. If $\frac{2}{9}$ of the non-Dingdong runners were girls, then $\frac{2}{9}$ of 18 = 4 girls, which leaves 14 boys. That means a total of 14 + 6 = 20 boys, out of a total of 30, or $\frac{20}{30} = \frac{2}{3}$. You will be happier if you make a tree chart:

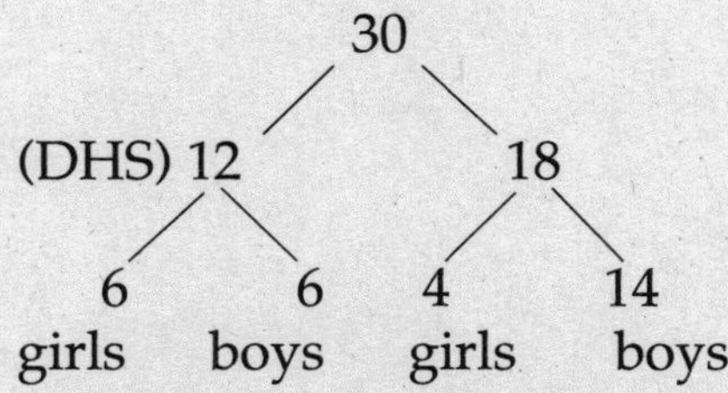

4. DECIMALS

To add, subtract, multiply, or divide decimals, use your calculator. Remember to check each number as you punch it in, and be extra careful with the decimal point.

To convert a fraction to a decimal, use your calculator to divide the numerator by the denominator:

$$\frac{1}{2} = 1 \div 2 = .5 \qquad \frac{5}{8} = 5 \div 8 = .625 \qquad \frac{4}{3} = 1.333$$

To convert a decimal to a fraction, count up the number of digits to the right of the decimal point, and put that many zeros in your denominator:

$$0.2 = \frac{2}{10} \qquad .314 = \frac{314}{1000} \qquad 2.23 = \frac{223}{100}$$

QUICK QUIZ #7

EASY

1 If $0.2p = 4$, then $4p =$

(A) 0.2
(B) 2
(C) 8
(D) 40
(E) 80

MEDIUM

10 If $z^2 = y^3$, and $y^2 = 16$, then $\frac{y}{z} =$

(A) 0.8
(B) 0.5
(C) 0.4
(D) 0.2
(E) 2

HARD

22 $\dfrac{\frac{ad}{bc}}{\frac{ac}{bd}} =$

(A) 1

(B) a^2c^2

(C) $\frac{a^2}{b^2}$

(D) $\frac{d^2}{c^2}$

(E) b^2d^2

Answers and Explanations: Quick Quiz #7

1 **E** If $0.2p = 4$, then $p = 20$, and $4p = 80$.

10 **B** If $y^2 = 16$, then $y = 4$. If $z^2 = 4^3$, then $z^2 = 64$ and $z = 8$.

So $\frac{y}{z} = \frac{4}{8} = \frac{1}{2} = .5$

22 **D** Remember that to divide fractions you flip the denominator and multiply. (Dividing is the same as multiplying by the reciprocal.)

$$\text{So } \frac{\frac{ad}{bc}}{\frac{ac}{bd}} = \frac{ad}{bc} \bullet \frac{bd}{ac} = \frac{adbd}{bcac} = \frac{abd^2}{abc^2} = \frac{d^2}{c^2}$$

---

5. PERCENTAGES

For some reason, many people get hung up on percents, probably because they are trying to remember a series of operations rather than using their common sense.

A percentage is simply a fractional part—50% of something is $\frac{1}{2}$ of something, and 47% is a little less than half. It is very helpful to approximate percents in this way, and not to think of them as abstract, meaningless numbers. 3.34% is very little of something, 0.0012% a tiny part of something, and 105% a little more than the whole.

Keep in mind that since percents are an expression of the fractional part, they do not represent actual numbers. If, for example, you're a salesperson, and you earn a 15% commission on what you sell, you'll get a lot richer selling Rolls-Royces than you will selling doughnuts. Even thousands of doughnuts. All 15%s are not created equal, unless they are 15% of the same number.

Now for the nitty-gritty:

To convert a percent to a decimal, move the decimal point two spaces to the left:

$50\% = .5$ $4\% = .04$ $.03\% = .0003$ $112\% = 1.12$

To convert a decimal to a percent, move the decimal point two spaces to the right:

$.5 = 50\%$ $.66 = 66\%$ $.01 = 1\%$ $4 = 400\%$

To convert a percent to a fraction, put the number over 100:

$50\% = \frac{50}{100}$ $4\% = \frac{4}{100}$ $106\% = \frac{106}{100}$ $x = \frac{x}{100}$

To get a percent of a number, multiply by the decimal. So to get 22% of 50, first change the percentage to a decimal by moving the decimal point 2 places to the left = .22. Then multiply on your calculator.

The second way to get a percent of a number is to transform your sentence into an equation. This is easier than it sounds. Convert the percent to a fraction and substitute × for *of*, = for *is*, and x for *what*.

What is 50% of 16?

transforms to $x = \frac{50}{100} \times 16$

This method is particularly useful for complicated percents:

What is 10% of 40% of 22?

transforms to $x = \frac{10}{100} \times \frac{40}{100} \times 22$

To calculate what percent one number is of another number, use the transformation method, substituting $\frac{x}{100}$ for *what percent.*

What percent of 16 is 8?

transforms to $\frac{x}{100} \times 16 = 8$

8 is what percent of 16?

transforms to $8 = \frac{x}{100} \times 16$

Notice that even though these equations look a little different, they will produce the same answer.

Another way: since figuring out what percentage one number is of another number is nothing more than figuring out the fractional part, you can solve these questions by setting up a proportion:

8 is what percent of 16?

can be written as $\frac{8}{16} = \frac{x}{100}$

What percent of 30 is 6?

can be written as $\frac{6}{30} = \frac{x}{100}$

What we're doing here is putting the part over the whole. Note that the part is usually the smaller number, but not always. The number that follows *of* will be the whole, and the number that comes right before or after *is* will be the part.

QUICK QUIZ #8

EASY

3 If 20% of p is 10, then 10% of p is

(A) 2
(B) 4
(C) 5
(D) 8
(E) 14

MEDIUM

11 Mabel agreed to pay the tax and tip for dinner at a restaurant with her four friends. Each of the friends paid an equal part of the cost of the dinner, which was $96. If the tax and tip together were 20% of the cost of the meal, Mabel paid how much less did than any one of her friends?

(A) $2.40
(B) $4.80
(C) $9.20
(D) $19.20
(E) $24.00

HARD

24 If 200% of 40% of x is equal to 40% of y, then x is what percent of y?

(A) 10%
(B) 20%
(C) 30%
(D) 50%
(E) 80%

Answers and Explanations: Quick Quiz #8

3 C 10% is half of 20%, and half of 10 is 5. That way we don't have to worry about p. To figure p, transform the sentence: $\frac{20}{100} \times p = 10$. $\frac{p}{5} = 10$, and $p = 50$. Now do the next step: $0.04 \times 50 = 5$.

11 B First calculate what each friend paid: $96 \div 4 = \$24$. Now do the percentage: $0.20 \times 96 = \$19.20$. Subtract the second number from the first. If you noticed that each of the four friends paid 25%, and Mabel paid 20%, you could take a fast shortcut by taking the difference, or 5% of 96. (If you picked D or E, you should reread the question before picking your final answer.)

24 D Plug in $100 = x$. 40% of 100 is 40, and 200% of 40 is $2 \times 40 = 80$. Now our question says that 80 is 40% of y, so $y = 200$. ($80 = 0.4y$) The question asks "x is what percent of y?" which you can write out as $100 = \frac{p}{100} \times 200$. Or you can set up a proportion: $\frac{p}{100} = \frac{100}{200}$. Or you can simply realize that 100 is half of 200, which is 50%.

MORE ON PERCENTAGES

To calculate percent increase, use the following formula:

difference or lower number = $\frac{x}{100}$

To calculate percent decrease, the formula changes slightly:

difference or higher number = $\frac{x}{100}$

QUICK QUIZ #9

MEDIUM

14 A store owner buys a pound of grapes for 80 cents and sells it for a dollar. What percent of the selling price of grapes is the store owner's profit?

(A) 10%
(B) 20%
(C) 25%
(D) 40%
(E) 80%

HARD

17 On the first test of the semester, Barbara scored a 60. On the last test of the semester, Barbara scored a 75. By what percent did Barbara's score improve?

(A) 12%
(B) 15%
(C) 18%
(D) 20%
(E) 25%

22 Randy's chain of used car dealerships sold 16,400 cars in 1989. If the chain sold 15,744 cars in 1990, by what percent did the number of cars sold decrease?

(A) 1%
(B) 4%
(C) 11%
(D) 40%
(E) 65%

Answers and Explanations: Quick Quiz #9

14 **B** First determine the store owner's profit. Change everthing to cents so that you're only working with one unit: 100 – 80 = 20. Now translate the question into math terms: $\frac{x}{100} \cdot 100 = 20$.

17 **E** Get the difference: 75 – 60 = 15. Put 15 over the lower number: $\frac{15}{60}$. Reduce the fraction to $\frac{1}{4}$, which is 25%. Or divide it on your calculator, which will give you 0.25. Convert to a percentage by moving the decimal 2 places to the right.

22 **B** Get the difference: 16,400 – 15,744 = 656. Put 656 over the higher number: $\frac{656}{16{,}400} = 0.04 = 4\%$. Use your calculator.

TIP: Estimating is always a good idea when you're doing a percentage question—a lot of the time there are silly answers that you can cross out before you do any math at all.

---

6. RATIOS

A ratio is like a percentage—it tells you how much you have of one thing compared to how much you have of another thing. For example, if you have hats and T-shirts in a ratio of 2:3, then for every two hats, you have three T-shirts. What we don't know is the actual number of each. It could be 2 hats and 3 T-shirts. Or it could be 4 hats and 6 T-shirts. Or 20 hats and 30 T-shirts.

A ratio describes a realtionship, not a total number.

To figure the actual numbers in a ratio, add the numbers in the ratio and divide into the total. Take that number and multiply by both parts.

> You have 24 CDs, jazz and rap, in a ratio of 1:2. How many do you have of each?

Add the numbers in the ratio: 1 + 2 = 3. Now divide 3 into 24 and get 8. Multiply that number by the parts in the ratio by that number: $8 \times 1 = 8$, and $8 \times 2 = 16$. So there are 8 jazz and 16 rap.

To check your answer, add the actual parts together. They should add up to the original total. 8 + 16 = 24, so the answer is right.

Note that the two actual numbers are really the same ratio, but unreduced. All we've done is multiply both parts of 1:2 by 8 to get 8:16.

TIP: Don't get the order of the ratio mixed up—if the problem says jazz and rap in a ratio of 1:2, the first number is jazz and the second is rap.

QUICK QUIZ #10

MEDIUM

12 If $\frac{x}{y} = \frac{4}{3}$ and $\frac{x}{k} = \frac{1}{2}$, then $\frac{k}{y} =$

(A) $\frac{1}{6}$

(B) $\frac{3}{8}$

(C) $\frac{2}{3}$

(D) $\frac{3}{2}$

(E) $\frac{8}{3}$

16 A teacher gave her class chocolate and vanilla ice cream bars in a ratio of 5:1. If x represents the number of students in her class, and $18 < x < 30$, how many students received chocolate ice cream bars, if each student received exactly one bar?

(A) 4
(B) 6
(C) 15
(D) 18
(E) 20

HARD

25 Forty gallons of a certain mixture is made up of 3 parts water and 5 parts Liquid *X*. If the mixture must be changed in order to be 75 percent Liquid *X*, how many gallons of Liquid *X* must be added to the original mixture?

(A) 15
(B) 20
(C) 35
(D) 50
(E) 60

Answers and Explanations: Quick Quiz #10

12 **E** Since $x = 4$ in one ratio and $x = 1$ in the other, we can't compare them. First we have to make them equal. If you multiply the second ratio by $\frac{4}{4}$, you get $\frac{4}{8}$. (Notice that if you multiply all parts of the ratio by the same number it doesn't change. It just takes an unreduced form.) Now the x's are the same in both ratios, so you can compare them, and $\frac{k}{y} = \frac{8}{3}$. You could also plug in, which would work at least as well.

16 **E** If the ratio is 5:1, we can add the parts and get 6 students—but 6 isn't between 18 and 30. Let's multiply the ratio by 3. Now we have 15:3—close, since that's a total of 18. Let's multiply by 4. Now we have 20:4. The parts of the ratio add up to 24, which is between 18 and 30, so it satisfies that part of the question. The ratio is chocolate: vanilla, so our answer is 20. (That was pretty hard for a medium question.)

25 **B** First figure out how much actual water and Liquid X you have. Add the parts of the ratio, $3 + 5 = 8$, and divide that sum into the total. $40 \div 8 = 5$. Now multiply the ratio by that quotient, making the ratio 15:25. (Same ratio, but representing actual gallons.) Now let's backsolve. Take B: if we add 20 gallons of Liquid X, we get 15:45. That's 45 gallons of X out of a total of 60 gallons, which is 75%.

7. PROPORTIONS

To set up a proportion, match categories on top and bottom. For example:

If 10 nails cost 4 cents, how much do 50 nails cost?

$$\begin{matrix}\text{(nails)} \\ \text{(cents)}\end{matrix}\ \frac{10}{4} = \frac{50}{x}$$

$$10x = 200, \text{ so } x = 20 \text{ cents}$$

The great thing about proportions is that it doesn't matter which is on top—if you match nails to nails and cents to cents (or whatever) you'll get the right answer. Be consistent.

To solve rate problems, set up a proportion:

If Bonzo rode his unicycle 30 miles in 5 hours, how long would it take him to ride 12 miles at the same rate?

$$\begin{matrix}\text{(miles)} \\ \text{(hours)}\end{matrix}\ \frac{30}{5} = \frac{12}{x}$$

$$30x = 60$$

$$x = 2 \text{ hours}$$

QUICK QUIZ # 11

EASY

2 Laura can solve 6 math questions in 12 minutes. Working at the same rate, how many minutes would it take Laura to solve 5 math questions?

(A) 6
(B) 8
(C) 9
(D) 10
(E) 11

MEDIUM

8 If 5 muffins cost \$3.00, how much do 2 muffins cost at the same rate?

(A) \$0.50
(B) \$0.75
(C) \$1.20
(D) \$1.50
(E) \$2.00

HARD

21 A factory produced 15 trucks of the same model. If the trucks had a combined weight of $34\frac{1}{2}$ tons, how much, in pounds, did one of the trucks weigh?
(One ton = 2000 pounds.)

(A) 460
(B) 2200
(C) 4500
(D) 4600
(E) 5400

Answers and Explanation: Quick Quiz #11

2 **D** $\frac{6}{12} = \frac{5}{x}$. Cross-multiply to get $6x = 60$, and $x = 10$.

8 **C** $\frac{5}{3.00} = \frac{2}{x}$, so $5x = 6.00$ and $x = 1.20$

21 **D** You can do this 2 ways: you can convert from tons to pounds first, or do it later. If you do it first, multiply 34.5×2000. That gives you 69,000. Your proportion should look like this: $\frac{15}{69{,}000} = \frac{1}{x}$. So $15x = 69{,}000$, and $x = 4600$. Or you can divide 15 into 34.5, which gives you 2.3 tons per truck. Then multiply 2.3 times 2000.

---

8. AVERAGES

You already know how to figure out an average. Of course you do. You can figure out your GPA, right?

To get the average of a set of numbers, add them up, then divide by the number of things in the set:

What's the average of 3, 5, and 10? $3 + 5 + 10 = 18$, and $18 \div 3 = 6$.

Most of the time on the SAT you are not given a set of numbers and asked for the average—they want to make it a little harder than that. There are three elements at work here: the sum of the numbers, the number of things in the set, and the average. To get any of these elements, you need to know the other two.

- The sum divided by the number of things = the average
- The sum divided by the average = the number of things
- The average multiplied by the number of things = the sum

TIP: Get the total, or sum of the numbers, and work from there. Usually you'll have to subtract something from the total. As always, avoid writing an equation.

BEWARE: Sometimes the average is called the *arithmetic mean* or the *mean*. They're all the same thing.

QUICK QUIZ #12

EASY

6 The average of 3 numbers is 22 and the smallest of these numbers is 2. If the other two numbers are equal, each of them is

(A) 22
(B) 30
(C) 32
(D) 40
(E) 64

MEDIUM

12 Four candy bars of different sizes cost at least $1.00 each. If it costs $2.50 to buy both of the 2 least expensive candy bars, what is the highest possible average cost of the other two candy bars?

(A) $0.75
(B) $1.00
(C) $1.25
(D) $1.50
(E) It cannot be determined from the information given.

HARD

23 In ten basketball games, Tommy shot an average of 8 foul shots per game and made an average of 40% of them. If there are four remaining games, and Tommy will shoot no more than 5 foul shots per game, what is the highest possible average foul-shooting percentage he can have for all 14 games?

(A) 45%
(B) 52%
(C) 65%
(D) 80%
(E) 140%

Answers and Explanations: Quick Quiz #12

6 C If the average of 3 numbers is 22, then their sum is 3 × 22 or 66. Take away the 2 and you've got 64 left. If the other 2 numbers are equal, divide 64 by 2 = 32.

12 E The total price of the 4 candy bars is at least $4.00. If we take away the price of the 2 least expensive bars, we get a cost of at least $1.50. But then the 2 most expensive bars could cost anything. They could cost a million dollars each for all we know. Or they could cost $2.00 and $3.00. Who knows? ("It cannot be determined from the information given" is a possible answer on a medium question, but usually not on a hard question. That's because the average person hopes you can't figure out hard questions, and the average person gets hard questions wrong.)

13 B Tommy shot a total of 80 foul shots in the first 10 games, and made 40% of them, and 0.4 × 80 = 32. If he shoots 5 shots per game in the 4 remaining games, that's 20 shots. Since we're looking for the highest average possible, let's say he makes all 20 of them. Now he has made 52 shots out of a total of 100, so his average percentage is 52%.

If you picked C, you forgot that he didn't make 52 shots out of 80, he made 52 shots out of 100. Remember to keep your eye on the total.

TIP: This would be a good question to estimate if you couldn't figure out how to do it. If Tommy does really well in the remaining 4 games, what's going to happen to his percentage? It will go up, but not too much, because he's already shot 40% in 10 games. So cross out D and E and guess.

---

9. MEDIAN, MODE, AND FREQUENCY

The main problem with median is that people get it confused with the average. They are similar, and sometimes they are equal, but only if your set of numbers advance in regular increments.

To find the median, first put the set of numbers in ascending order. If the set has an odd number of elements, the median is the middle number.

set: 1, 4, 9, 18, 54 median: 9
set: 2, 4, 4, 4, 5 median: 4

If the set has an even number of elements, the median is *the average of* the two middle numbers.

set: 3, 15, 17, 74 median: 16
set: 1, 6, 7, 8 median: 6.5

To find the mode, just look to see which number in the set appears the most often.

set: 1, 1, 3, 5, 3, 4, 22, 3, 6 mode: 3

If two numbers appear the most often, they are *both* the mode.

set: 2, 3, 5, 7, 7, 2 mode: 2 and 7

To find the frequency, list the numbers of the set vertically, in ascending order. Then count up the number of times each one appears. Frequency is like mode, only you figure out how often *every* number appears in the set.

Set A: { 2, 1, 19, 4, 5, 1, 3, 5, 19, 22, 2, 3, 4, 1}

#	Frequency
1	3
2	2
3	2
4	2
5	2
19	2
22	1

Note: the numbers in the frequency column should add up to the total of numbers in the set.

QUICK QUIZ #13

EASY

Set Q: {10, 2, 3, 5, 1, 7, 5, 2}

6 If the smallest and largest numbers in Set Q are removed, what is the median of Set Q?

(A) $3\frac{1}{2}$
(B) 4
(C) 5
(D) 6
(E) 7

MEDIUM

August Temperature Readings in Plainville

Temperature Range (in degrees)	Frequency (in days)
66–70	3
71–75	6
76–80	5
81–85	8
86–90	7
91–95	2

16 The difference between the temperature range of the mode and the lowest possible recorded temperature falls within which of the following ranges?

(A) 11°–15°
(B) 13°–17°
(C) 15°–17°
(D) 15°–19°
(E) 19°–22°

HARD

24 A group of 12 students received the following grades: A, C, D, B, C, A, D, C, B, C, B, F, with A being the highest grade and F the lowest. The grades were then assigned number values: A = 4, B = 3, C = 2, D = 1, and F = 0. If the teacher announced that every student had the option to receive the grade that equaled the number value of the mode minus the frequency of grade F, how many students would benefit from the teacher's offer?

(A) 0
(B) 1
(C) 4
(D) 5
(E) 7

Answers and Explanations: Quick Quiz #13

3 **B** Take out 10 and 1. Now write the numbers down in order: 2, 2, 3, 5, 5, 7. The middle of the list falls between 3 and 5, so the median is 4.

16 **D** The mode is 81 – 85, since there are more days with temperatures in that range than any other. The lowest possible recorded temperature is 66, and to get the difference between that and the mode, we need to subtract 66 from 81 and from 85. That should give you 15 and 19. (Pretty hard for a medium question.)

24 **B** The mode is C, which has a value of 2. The frequency of F is 1. 2 – 1 = 1, which is the number value for grade D. Only students with a grade lower than D will benefit, so that leaves you with the 1 student who got an F.

This question is easier than it looks. When you're dealing with a complicated word problem, don't freak out. Just take it one step at a time, and reread the question several times.

10. EXPONENTS

An exponent tells you how many times to multiply a number by itself. So x^3 is really shorthand for $x \cdot x \cdot x$. If you have a momentary lapse and can't remember the following rules, it may help to write your problem out the long way and work from there.

For exponents with the same base:
To multiply, add the exponents: $x^2 \times x^5 = x^{2+5} = x^7$
To divide, subtract the exponents: $x^6 \div x^3 = x^{6-3} = x^3$
To raise the power, multiply: $(x^4)^3 = x^{4 \cdot 3} = x^{12}$
You cannot add or subtract, so $x^6 + x^3$ is just $x^6 + x^3$. You can't reduce it.

For exponents with different bases:
Forget it. You can't do anything. Don't try to multiply, divide, add, subtract, or cancel.

To deal with exponents and parentheses, remember that the exponent carries over to all parts within the parentheses:

$$2(3a^3)^2 = 2[(3^2)(a^6)] = 2(9a^6) = 18a^6$$

Keep in mind that 1 raised to any power is still just 1. ($1^{357} = 1$.)
Negative numbers with even exponents are positive; negative numbers with odd exponents are negative. Fractions with exponents get smaller, not bigger.

QUICK QUIZ #14

EASY

1 If $(3x)^2 = 81$, then $x =$

(A) 2
(B) 3
(C) 6
(D) 9
(E) 12

MEDIUM

16 If $a > 0$, $b < 1$, and $c < 0$, which of the following must be true?

(A) abc is positive

(B) abc is negative

(C) a^2b^2 is positive

(D) ab^2c^2 is positive

(E) $a^3b^3c^3$ is negative

HARD

20 $a^7 \bullet b^7 =$

(A) $(ab)^7$
(B) $(a+b)^7$
(C) $(ab)^{14}$
(D) $(a+b)^{14}$
(E) $(ab)^{49}$

Answers and Explanations: Quick Quiz #14

1 **B** Square everything within the parentheses, so you get $3^2x^2 = 81$, or $9x^2 = 81$. Divide through by 9 and you get $x^2 = 9$ and $x = 3$.

16 **D** You life will be easier if you make a little chart showing the signs of each variable:

a +

b ?

c –

Now go to the answers. If we don't know the sign of *b*, we don't know the sign of choices A, B, or E. In C, a^2 is positive, b^2 will have to be positive no matter what the sign of *b* is, and *c* is negative. So the whole thing is negative. Our answer is D.

20 **A** If you don't remember the rules for exponents, you can attack exponent problems by expanding out the terms:
$a^7 \bullet b^7 = a \bullet a \bullet a \bullet a \bullet a \bullet a \bullet a \bullet b \bullet b \bullet b \bullet b \bullet b \bullet b \bullet b =$
Now rearrange the terms:
$a \bullet b \bullet a \bullet b \bullet a \bullet b \bullet a \bullet b \bullet a \bullet b \bullet a \bullet b \bullet a \bullet b$. By bracketing off each $(a \bullet b)$ pair, you can see the $a^7 \bullet b^7 = (a \bullet b)$ times itself 7 times. Plugging in also works. Let $a = 2$ and $b = 3$. Use your calculator.

11. ROOTS

In the seventies, it was a wildly popular mini-series. On the SAT, it's a type of question that gives a lot of people a big headache. Expect to see about 1-4 square root questions on your test.

A square root is just a backwards exponent; in other words, the number under the $\sqrt{\ }$ is what you get when you raise a number to a power of 2.

$\sqrt{4} = 2$ $\sqrt{36} = 6$ $\sqrt{1} = 1$

To multiply or divide square roots, just multiply or divide as usual.

$\sqrt{7} \bullet \sqrt{3} = \sqrt{21}$ $\sqrt{15} \div \sqrt{3} = \sqrt{5}$

To add or subtract square roots, first make sure you have the same number under the $\sqrt{\ }$. Then add or subtract the number outside of the $\sqrt{\ }$.

$5\sqrt{3} + 2\sqrt{3} = 7\sqrt{3}$ $6\sqrt{2} - \sqrt{2} = 5\sqrt{2}$

Note that:

- A square root multiplied by itself is just that number without the $\sqrt{\ }$. ($\sqrt{3} \bullet \sqrt{3} = 3$)
- The square root of a fraction gets bigger. For example, $\sqrt{\frac{1}{4}} = \frac{1}{2}$.
- The square root of a number is always positive (on the SAT, anyway).
- The square root of 1 is 1.

QUICK QUIZ #15

EASY

5 What is the reciprocal of $\sqrt{\frac{1}{4}}$?

(A) $\frac{1}{4}$

(B) $\frac{1}{2}$

(C) 2

(D) 4

(E) 16

MEDIUM

15 $\frac{1}{4}$ is what percent of $\left(\sqrt{5}\right)^3$?

(A) .25

(B) $\frac{5}{\sqrt{5}}$

(C) 5

(D) $5\sqrt{50}$

(E) 50

HARD

25 $\frac{\sqrt{6x} \bullet \sqrt{21x} \bullet \sqrt{7x}}{\sqrt{2x}} =$

(A) 3x
(B) $7\sqrt{x}$
(C) 21x
(D) $21\sqrt{x}$
(E) $7\sqrt{2x}$

Answers and Explanations: Quick Quiz #15

5 C $\sqrt{\frac{1}{4}} = \frac{1}{2}$, and the reciprocal of $\frac{1}{2}$ is 2. If you picked B, you didn't finish the question.

15 B $\left(\sqrt{5}\right)^3 = \sqrt{5} \bullet \sqrt{5} \bullet \sqrt{5} = 5\sqrt{5}$. Even better, use your calculator.

Now set up a proportion: $\frac{\frac{1}{4}}{5\sqrt{5}} = \frac{x}{100}$. So $x = \frac{25}{5\sqrt{5}} = \frac{5}{\sqrt{5}}$.

25 C Remember that you can multiply everything under the square root signs. Breaking down into like terms first gives you:

$$\frac{\sqrt{2x} \bullet \sqrt{3} \bullet \sqrt{3} \bullet \sqrt{7x} \bullet \sqrt{7x}}{\sqrt{2x}}$$

$$\left(\sqrt{3} \bullet \sqrt{3}\right) \bullet \left(\sqrt{7x} \bullet \sqrt{7x}\right) = 3 \bullet 7x = 21x$$

---

12. PROBABILITY AND COMBINATIONS

There won't be zillions of these on the SAT, but you may as well get them right if they show up.

Here's an example:

> If you flip a coin twice, what's the probability of getting 2 heads?

Solution: First write out all the possibilities:

H – H
H – T
T – T
T – H

That's a total of 4 possibilities, and only 1 is heads-heads. The probability is 1 out of 4, or $\frac{1}{4}$.

Another example:

> Mr. Jones must choose four of the following five flavors of jellybean: apple, berry, coconut, kumquat, and lemon. How many different combinations of flavors can Mr. Jones choose?

Solution: Again, write out the possibilities, being careful to go in order so you don't leave anything out. Use the first letter of the flavors as a kind of shorthand:

Flavors: ABCKL
Possibilities: ABCK
ABCL
ABKL
ACKL
BCKL

So there are 5 possible combinations. Notice that we only went in one direction to get our combinations—the order doesn't matter, and if you go forward and backward (KCAB) you'll get impossibly confused. Being methodical is the key here.

QUICK QUIZ #16

EASY

5 What is the probability of randomly choosing a white marble from a bag that contains 4 white marbles, 2 blue marbles, and 3 green marbles?

(A) $\frac{1}{4}$ (B) $\frac{2}{5}$ (C) $\frac{2}{7}$ (D) $\frac{4}{9}$ (E) $\frac{4}{5}$

MEDIUM

Column A	Column B
steak	french fries
hamburger	spinach
fish	peas
pork chops	salad
	rice

9 If a diner at a certain restaurant may choose 1 item from Column A and 1 item from Column B, how many different combinations may the diner choose?

(A) 5 (B) 9 (C) 15 (D) 20 (E) 25

HARD

25 The probability of randomly picking a chocolate cookie out of a certain bag is 1 out of 6, and the probability of picking a lemon cookie is 2 out of 9. If the bag contains only chocolate, vanilla, and lemon cookies, how many vanilla cookies could be in the bag?

(A) 6 (B) 7 (C) 11 (D) 18 (E) 20

Answers and Explanations: Quick Quiz #16

5 **D** The total number of marbles is 9, and 4 of them are white. That means there's a 4-in-9 chance of picking a white marble. Keep in mind that the total goes on the bottom and the part goes on the top.

$\frac{9}{4}$ isn't one of the choices here, but a lot of people might have wanted to pick it.

9 **D** There's a nice shortcut to this kind of combination question: just add up the total for both columns and multiply. Column A has 4 items and Column B has 5. $5 \cdot 4 = 20$. Or you could write out all the combinations, but that would take a while.

25 **C** If you add the fractions together, you get $\frac{7}{18}$, which is the probability of picking chocolate or lemon. So let's just say there are 18 cookies total, and the other 11 are vanilla.

CHAPTER 3

Algebra

ALGEBRA

In the section on strategy, we gave you some ways to avoid algebra altogether—but you still need to be able to work with simple equations and review some other algebraic principles that don't exactly crop up in everyday life.

1. SIMPLE EQUATIONS

Sometimes you can backsolve these, sometimes not. You will definitely need to be comfortable manipulating equations to do well on the SAT.

To solve a simple equation, get the variable on one side of the equals sign and the numbers on the other.

$$\begin{aligned} 9x - 4 &= 12 + x \\ 8x &= 16 \\ x &= 2 \end{aligned}$$

We just added 4 to both sides and subtracted x from both sides. Then we divided both sides by 8. You can add, subtract, multiply or divide either side of an equation, but remember that what you do to one side you have to do to the other.

To solve a proportion, cross-multiply:

$$\frac{3}{x} = \frac{1}{2}$$

$$x = 6$$

Note that you can't cancel across an equals sign!

QUICK QUIZ #17

EASY

3 If $\frac{3x}{5} = \frac{x+2}{3}$, what is the value of x?

(A) $\frac{1}{2}$

(B) 1

(C) 2

(D) $2\frac{1}{2}$

(E) 3

MEDIUM

14 If $\frac{5}{x} = \frac{y}{10}$ and $x - y = y$, then $y + x =$

(A) 5
(B) 10
(C) 15
(D) 25
(E) 50

HARD

21 If $\frac{a+1}{b+1} = \frac{a}{b}$, then $(a + b)(a - b) =$

(A) –1
(B) 0
(C) 1
(D) 2
(E) It cannot be determined from the information given.

Answers and Explanations: Quick Quiz #17

3 D Cross-multiply, and you get $9x = 5(x + 2)$

$$9x = 5x + 10$$
$$4x = 10$$
$$x = 2\frac{1}{2}$$

14 C Plug 10 for x and 5 for y. Both equations are satisfied by those numbers. So $y + x = 15$.

Just to show you the kind of algebra that you'd be forced to do if you didn't plug in—first, cross-multiply to get $xy = 50$. Your other equation is $x - y = y$, so $x = 2y$. Substitute that x into the first equation, and you get $2y^2 = 50$, or $y^2 = 25$. So $y = 5$. Substitute $y = 5$ into either equation and solve for x. You get $x = 10$. Now add 'em up and you get $x + y = 15$. Lots more work, huh? If you don't plug in when you can, it's really going to slow you down. And that's the least of it. You're also more likely to get the question wrong, because the algebra takes so many steps.

21 B Sometimes you can't plug in, because it's too hard to find numbers that satisfy the equation. So this time we have to do the algebra. First, cross-multiply: $ab + a = ab + b$, so $a = b$. The question asks for $(a + b)(a - b)$. *Now* plug in. If $a = 2$ and $b = 2$, then $(a - b)$ is 0 and the whole thing is 0.

Oh—did you pick E? Never pick "It cannot be determined" if you're dazed and confused. Not on a hard question. That answer is correct when you can get more than one right answer, and that almost always happens on medium questions.

---

2. QUADRATIC EQUATIONS

Even the name is scary—what does it mean, anyway? No matter. All you need to know are a few simple things: factoring and recognizing perfect squares.

To factor, first draw a pair of empty parentheses. Deal with the first term, then the signs, then the last term. For example:

$x^2 + x - 12$ $(\quad)(\quad)$
$(x\quad)(x\quad)$ ....first term
$(x +\quad)(x -\quad)$ ....signs
$(x + 4)(x - 3)$ ....last term

Check your factoring by multiplying the terms:

first term = $x \times x = x^2$
inner term = $4x$
outer term = $-3x$
last term = $4 \times -3 = -12$

Then add them up:

$$x^2 + 4x + -3x + -12 = x^2 + x - 12$$

Some guidelines:
If the last term is positive, your signs will be either +, + or –, –
If the last term is negative, your signs will be +, –
Your first try may not be right—don't be afraid to mess around with it a little.

To recognize the difference of two squares, memorize the following:

$$(x + y)(x - y) = x^2 - y^2$$

This format works whether you have variables, as above, or numbers:

$$57^2 - 43^2 = (57 + 43)(57 - 43) = 100 \times 14 = 1400$$

One more thing—memorize the following:

$$(x + y)(x + y) = x^2 + 2xy + y^2$$

TIP: When you see either of the above—$x^2 - y^2$ or $(x + y)^2$—just convert it to its other form. That should lead you straight to the correct answer.

QUICK QUIZ #18

EASY

7 If $\frac{x^2+5x+6}{x+2} = 12$, then $x =$

(A) -2
(B) 2
(C) 3
(D) 6
(E) 9

MEDIUM

15 If $a - b = 3$ and $a^2 - b^2 = 21$, then $a =$

(A) -3
(B) -2
(C) 2
(D) 5
(E) 7

HARD

20 If $x < 0$ and $(2x - 1)^2 = 25$, then $x^2 =$

(A) -4
(B) -2
(C) 3
(D) 4
(E) 9

Answers and Explanations: Quick Quiz #18

4 E First factor the expression to $(x + 3)(x + 2)$. Now you have $\frac{(x+3)(x+2)}{x+2} = 12$. The $(x + 2)$ cancels, and you have $x + 3 = 12$, so $x = 9$. Or you could backsolve: if $x = 9$, $\frac{9^2 + 5(9) + 6}{9+2} = 12$, or $\frac{132}{11} = 12$. Looks funny, but it works.

15 D Factor $a^2 - b^2$ to equal $(a + b)(a - b) = 21$. If $a - b = 3$, then $a + b = 7$. Here you could do one of 2 things. You can try some different numbers and see what satisfies both simple equations, or you could add the 2 equations together and get $2a = 10$, $a = 5$.

20 D Lots of algebra:

$$(2x-1)^2 = 25$$

$$(2x-1)(2x-1) = 25$$

$$4x^2 - 4x + 1 = 25$$

$$4x^2 - 4x - 24 = 0$$

$$x^2 - x - 6 = 0$$

$$(x - 3)(x + 2) = 0$$

So x can be 3 or -2. If x is negative, it has to be -2, and $-2^2 = 4$. You could also backsolve, but you have to remember that the question asks for x^2, not x. That means D and E are good answers to try, since they're squares.

Don't forget that one of your main jobs on the SAT is following directions. If you picked B or C, we suspect you did most of the problem correctly but forgot that x is negative, or failed to square x. Don't let carelessness rob you of your hard-earned points!

3. SIMULTANEOUS EQUATIONS

Two different equations, two different variables. You will not usually have to solve for both variables.

To solve simultaneous equations, stack 'em up, and either add or subtract:

If $2x + 3y = 12$ and $3x - 3y = -2$, what is the value of x?

$$\begin{array}{r} 2x + 3y = 12 \\ +\ \ 3x - 3y = -2 \\ \hline 5x = 10 \\ x = 2 \end{array}$$

If we had subtracted, we'd have gotten $-x + 6y = 14$, which wouldn't get us anywhere. If you choose the wrong operation, no big deal, just try the other one.

TIP: Don't automatically start solving for x and y—you may not need to. Focus on what the question is specifically asking for.

QUICK QUIZ #19

MEDIUM

9 If $3x + 3y = 4$ and $2x - 3y = 1$, what is the value of x?

(A) $\frac{1}{3}$

(B) 1

(C) 3

(D) 5

(E) 6

10 All of the items on a breakfast menu cost the same whether ordered with something else or alone. Two orders of pancakes and one order of bacon costs $4.92. If two orders of bacon cost $3.96, what does one order of pancakes cost?

(A) $.96
(B) $1.47
(C) $1.98
(D) $2.94
(E) $3.20

11 If $3x + 5y = 15$ and $x - 2y = 10$, then $2x + 7y =$

(A) 5
(B) 10
(C) 15
(D) 25
(E) 50

Answers and Explanations: Quick Quiz #19

9 **B** Stack 'em and add:

$$\begin{array}{r} 3x + 3y = 4 \\ +\ \ 2x - 3y = 1 \\ \hline 5x \qquad = 5 \\ x = 1 \end{array}$$

10 **B** If 2 orders of bacon cost \$3.96, divide by 2 to get the cost of 1 order of bacon. That's \$1.98. If 2 orders of pancakes and 1 order of bacon costs \$4.92, we can subtract the \$1.98 and get \$2.94, which is cost of 2 orders of pancakes. Divide by 2 to get 1 order of pancakes. That's \$1.47.

The simultaneous equations would be $2p + b = 4.92$ and $2b = 3.96$. Solve for b and then use that number to solve for p in the first equation. It's even possible to backsolve this question, but the numbers are so irregular and hard to work with, it probably isn't worth it.

11 **A** Stack 'em and subtract:

$$\begin{array}{r} 3x + 5y = 10 \\ -\ \ (x - 2y = 10) \\ \hline \end{array} \qquad \begin{array}{r} 3x + 5y = 15 \\ -\ \ x + 2y = -10 \\ \hline 2x + 7y = 10 \end{array}$$

That's it. You don't have to solve for x or y individually. Less work is good. (Be careful with the signs when you subtract one equation from another.)

---

4. INEQUALITIES

Treat these just like equations, but remember one rule: **if you multiply or divide through by a negative number, the sign changes direction.**

$x + 6 > 10$	$2x > 16$	$-2x > 16$
$x > 4$	$x > 8$	$x < -8$

BEWARE: It's very easy to mix up the direction of the $>$ or $<$ sign. Be extra careful.

QUICK QUIZ #20

EASY

3 If $3x + 7 < 5x - 4$, then

(A) $\frac{11}{2} < x$

(B) $x < \frac{3}{2}$

(C) $x < \frac{11}{8}$

(D) $x > \frac{2}{3}$

(E) $\frac{11}{2} > x$

MEDIUM

11 If $3b + 8 > 6 + 2b$, and b is a negative integer, then $b =$

(A) 1
(B) 0
(C) –1
(D) –2
(E) –3

HARD

20 If $3t - 3 > 6s + 9$ and $t - 5s < 12$, and s is a positive integer less than 4, then t could be any of the following EXCEPT

(A) 6
(B) 8
(C) 10
(D) 12
(E) 23

Answers and Explanations: Quick Quiz #20

3 A Treat the inequality just like an equation—subtract $3x$ from both sides, and you get $7 < 2x - 4$. Add 4 to both sides, and you get $11 < 2x$. Divide through by 2, which leaves you with $\frac{11}{2} < x$.

11 C Move the b's to one side and the integers to the other, and you get $b > -2$. If b is a negative integer, the only possibility is -1.

20 A Manipulate the inequalities so that they're more manageable. The first one can be reduced to $t - 1 > 2s + 3$, or $t > 2s + 4$. The second inequality can be manipulated to $t < 12 + 5s$. Now let's plug in. If s is a positive integer less than 4, start with the first possible value for s, which is $s = 1$. Plugged into both inequalities, that gives you $t > 6$ and $t < 17$. Cross out B, C, and D. Now try the highest possible value for s, which is $s = 3$. That gives you $t > 12$ and $t < 25$. Cross out E, and pick the answer that's left: A.

---

5. FUNCTIONS

Forget the $f(x)$ functions you had in algebra class—they aren't on the SAT. These funny-looking problems use strange symbols that give some sort of direction. Fear not. Think of them as "Simon Says," and just follow the directions. For example:

> For any integer t, $[t] = t = t^2 - t$. What is the value of $[4] - [3]$?

Solution: Let's take [4] first. The direction tells us to square the number, and then add the number, so $4^2 + 4 = 20$. Now let's do the same for [3]. $3^2 + 3 = 12$. So $[4] - [3] = 20 - 12 = 8$.

QUICK QUIZ #21

MEDIUM

10 For all positive integers x, $x♠ = x^2 - 2$. Which of the following is equal to $3♠ + 2$?

(A) 1
(B) 5
(C) 9
(D) 11
(E) 25

16 If $[a + b] = a^2 - b^2$, then $\frac{[x+y]}{x+y} =$

(A) $x + y$
(B) $x - y$
(C) $2x - 2y$
(D) 1
(E) $(x + y)^2$

HARD

25 For all numbers y, $<y> = \frac{y-1}{y+1}$. Which of the following is equal to the positive reciprocal of <0>?

(A) <0>
(B) <1> + <1>
(C) <3> – <2>
(D) <2> + <5>
(E) <4> + <3>

Answers and Explanations: Quick Quiz #21

10 C Just follow the directions, which are to square the number and subtract 2. So $3♠ = 3^2 - 2 = 7$. $2♠ = 2^2 - 2 = 2$, and $7 + 2 = 9$.

16 **B** $[x + y] = x^2 - y^2$, which factors to $(x + y)(x - y)$. When you divide, the $(x + y)$ term cancels, and you're left with $x - y$.

25 **D** Hard function questions can get really gnarly. First of all, $\langle 0 \rangle = \frac{0-1}{0+1} = -\frac{1}{1} = -1$. The positive reciprocal of –1 is 1. Now we have to functionize the answer choices as well. (Don't be fooled into picking C. That's too easy for Question 25.) In D, $\langle 2 \rangle = \frac{2-1}{2+1} = \frac{1}{3}$. $\langle 5 \rangle = \frac{5-1}{5+1} = \frac{4}{6} = \frac{2}{3}$. And $\frac{1}{3} + \frac{2}{3} = 1$. If you tried D first, you're a lucky dog. In a question like this, A and C look kind of appealing, so don't try them first. They're traps. Other than that, you just have to slog through the other answers until you hit the right one.

That's it for algebra. Be sure to review plugging in and backsolving in the Strategy section, because these two techniques work best for most algebra problems.

CHAPTER 4

Geometry

GEOMETRY

You're not going to believe how simple this is—no proofs, no trig, no parabolas. Just a few rules, a couple of formulas, and your common sense. And don't forget about estimating.

DEFINITIONS

quadrilateral	any 4-sided figure
isosceles	a triangle with 2 equal sides and 2 equal angles
equilateral	a triangle with 3 equal sides, therefore 3 equal angles (60°each)
bisect	cut in 2 equal parts
parallel	lines that will never intersect (think railroad tracks)
perpendicular	2 lines that intersect to form 90° angles
hypotenuse	the longest leg of a right triangle, opposite the right angle
radius	a line from the center of a circle to the edge of the circle (half the diameter)
diameter	a line directly through the center of a circle—the longest line you can draw in a circle
circumference	the distance around a circle
arc	part of a circumference
chord	a line that intersects a circle (it's shorter than the diameter)
diagonal	a line from one corner of a square to its opposite corner
equidistant	exactly in the middle
perimeter	the distance around a figure (linear)
area	the space inside a figure (2-dimensional)
volume	the space inside a figure (3-dimensional)

1. LINES AND ANGLES

A line has 180°, so the angles formed by any cut to your line will add up to 180:

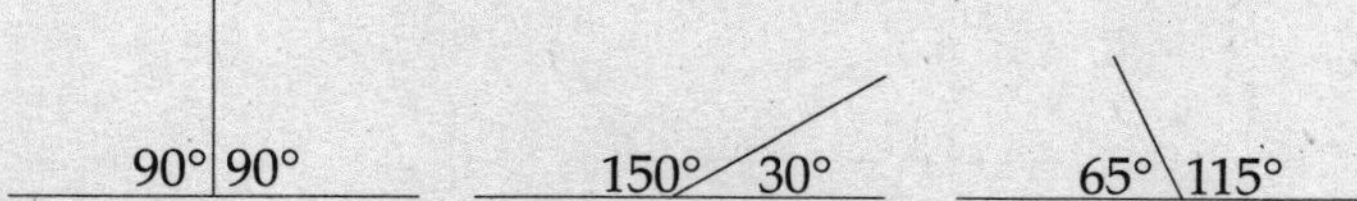

Two intersecting lines form a pair of **vertical angles**, which are equal:

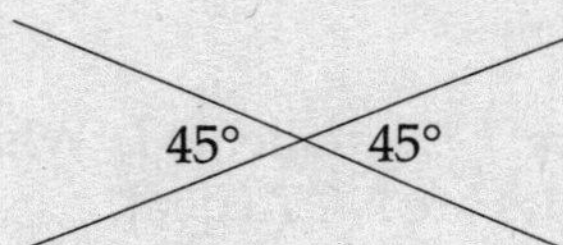

Parallel lines cut by a third line will form two kinds of angles, big ones and little ones. All the big ones are equal to each other, all the little ones are equal to each other. Any big angle plus any little angle will equal 180°:

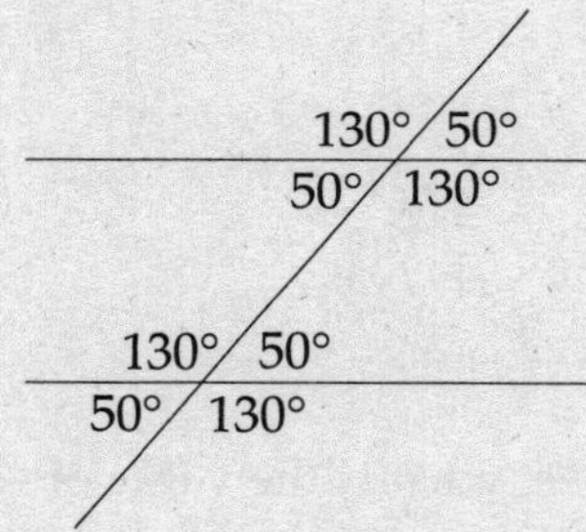

QUICK QUIZ #22

EASY

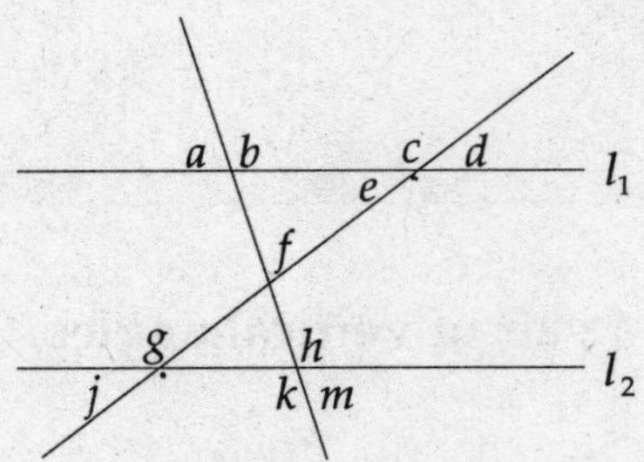

4 In the figure above, l_1 is parallel to l_2. Which of the following angles are NOT equal?

(A) c and g
(B) b and h
(C) a and m
(D) a and k
(E) d and j

MEDIUM

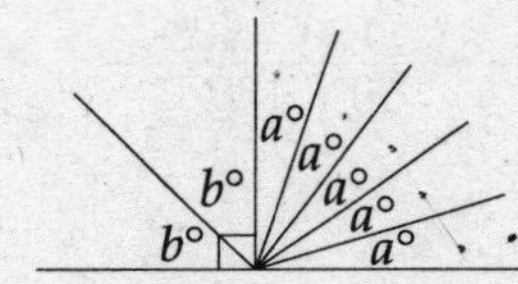

10 In the figure above, what is the value of $4a - b$?

(A) 18°
(B) 27°
(C) 45°
(D) 54°
(E) 115°

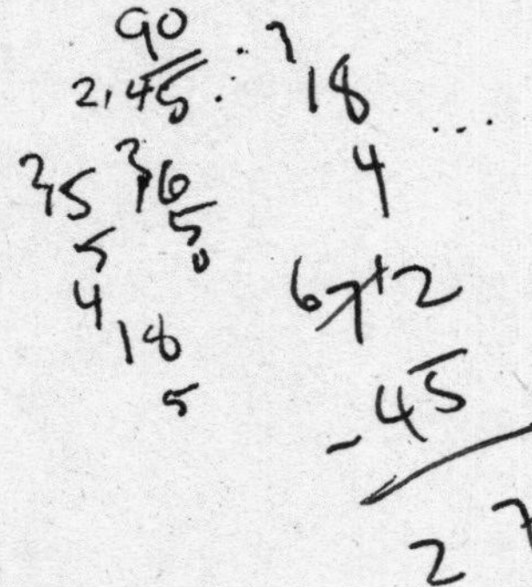

HARD

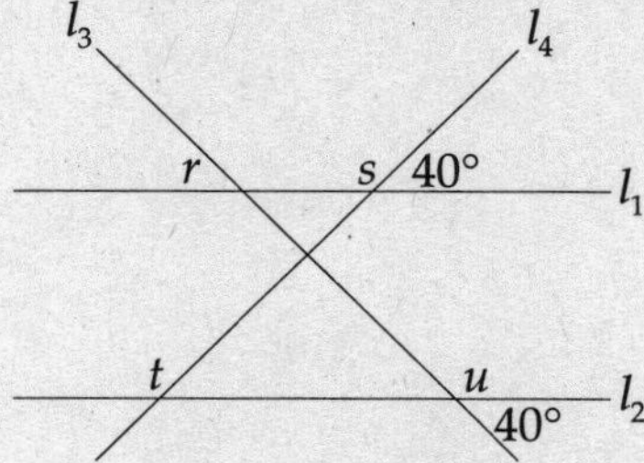

Note: Figure not drawn to scale.

18 Which of the following must be true?

(A) $l_1 \parallel l_2$
(B) l_3 bisects l_4
(C) $r = 40°$
(D) $s = t$
(E) $u = 140°$

Answers and Explanations: Quick Quiz #22

4 **D** Start with A and cross off as you go along. In D, $a = m$, not k. Keep in mind that the 2 lines cutting through l_1 and l_2 aren't parallel, and so the angles made by one line have no relationship to the angles made by the other line.

10 **B** Estimate first. Outline the measurement of 4 of the a's. That's about 60? Now pretend you are subtracting b, about 45. How much is left? Not so much, right? Cross out D and E. Now do the math: $2b = 90°$, so $b = 45°$. $5a = 90$, so $a = 18°$. Now plug those numbers into the equation: $4(18) - 45 = 27$.

18 **E** This question is actually very easy, as long as you don't pick the first answer that looks halfway decent and not even get to E. Angle u has to be 140°, because it's on a straight line with the angle marked 40°. All of the other answers look like they're true, but they are only maybe true. The only thing you know for sure is that angles on the same line add up to 180°, and vertical angles are equal. None of these lines are necessarily parallel, so you can't assume anything else.

2. TRIANGLES

Triangles have 180°.

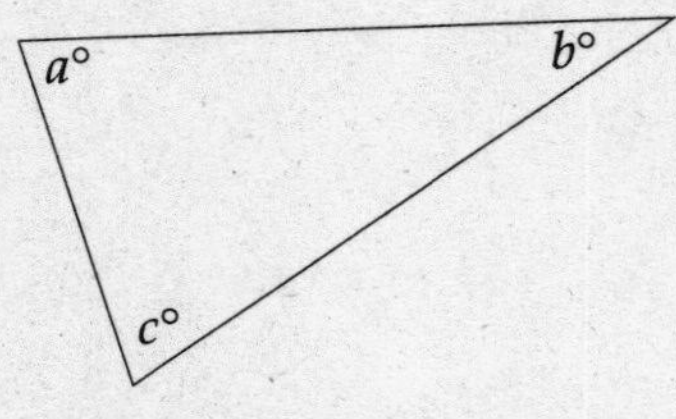

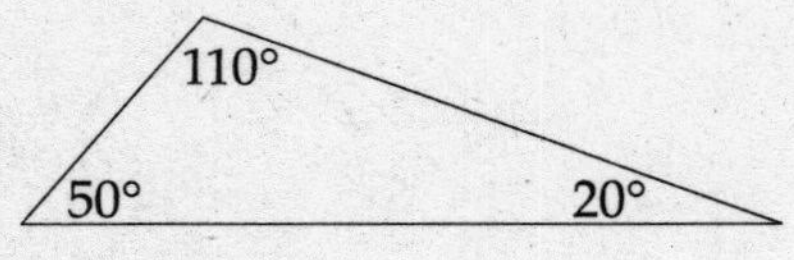

$a + b + c = 180°$

$50° + 20° + 110° = 180°$

Area = $\frac{1}{2}bh$.

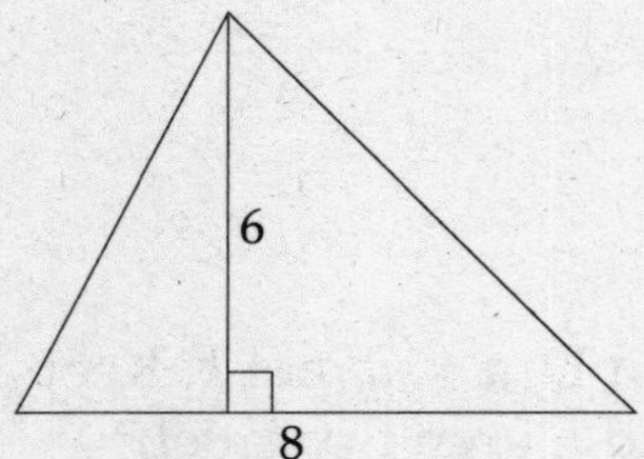

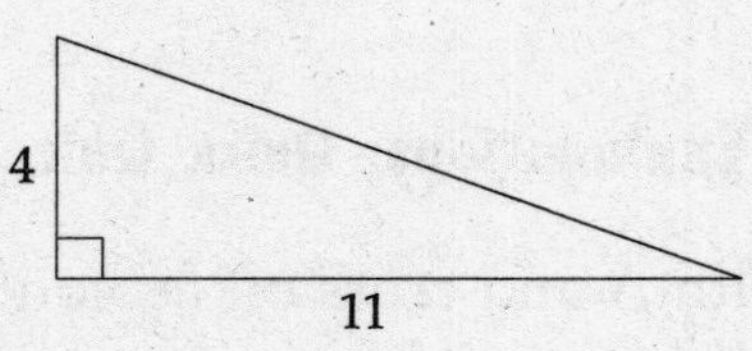

area = $\frac{1}{2}(8)(6) = 24$

area = $\frac{1}{2}(11)(4) = 22$

Perimeter: add up the sides.

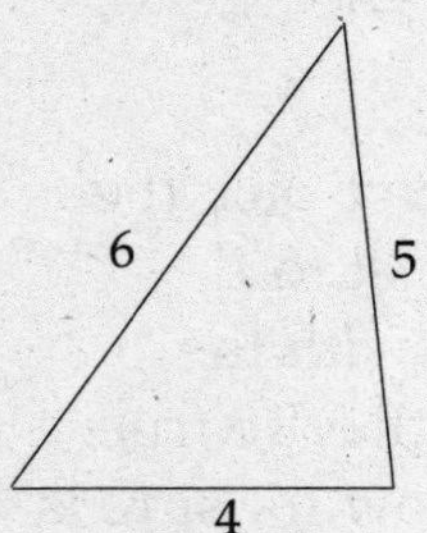

Perimeter = 15

Right triangles have a right, or 90°, angle:

Isosceles triangles have 2 equal sides and 2 equal angles:

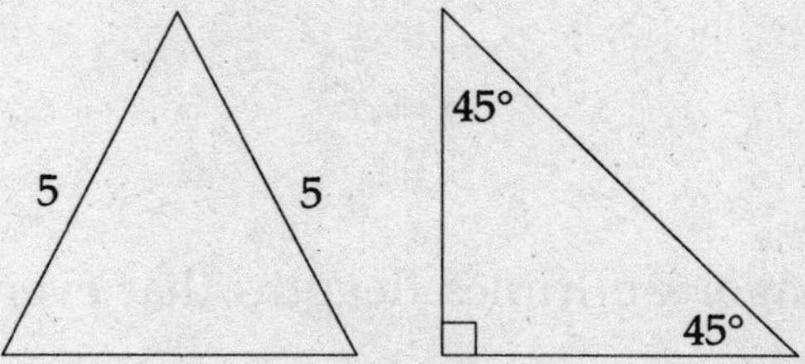

Equilateral triangles have 3 equal sides and 3 equal angles:

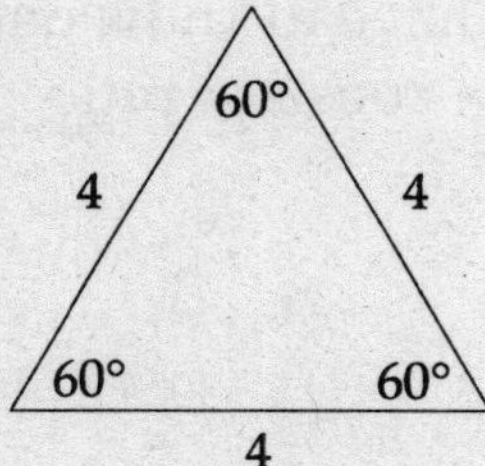

Similar triangles have equal angles and proportional sides:

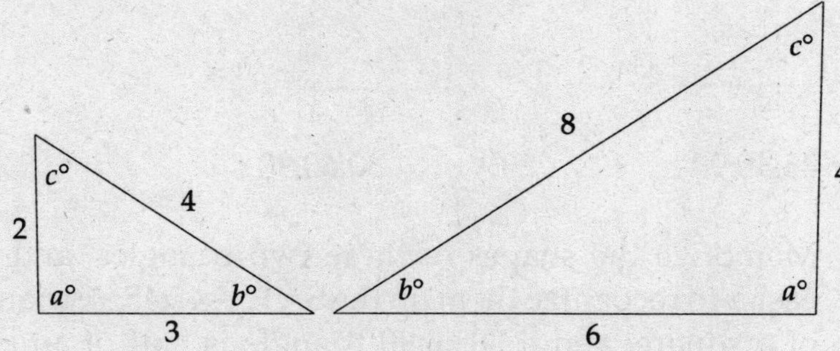

THE WONDERFUL WORLD OF RIGHT TRIANGLES

For any right triangle, if you know the lengths of two of the sides, you can figure out the length of the third side by using the Pythagorean Theorem:

$$a^2 + b^2 = c^2$$

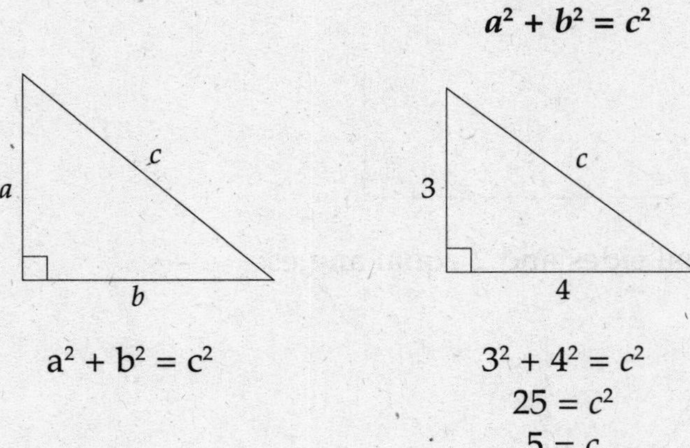

$a^2 + b^2 = c^2$

$$3^2 + 4^2 = c^2$$
$$25 = c^2$$
$$5 = c$$

Some common Pythagorean triples (lengths that evenly satisfy the theorem) are:

3:4:5 6:8:10 5:12:13

In two special cases, you only have to know one side to figure out the other two, because the sides are in a constant ratio.

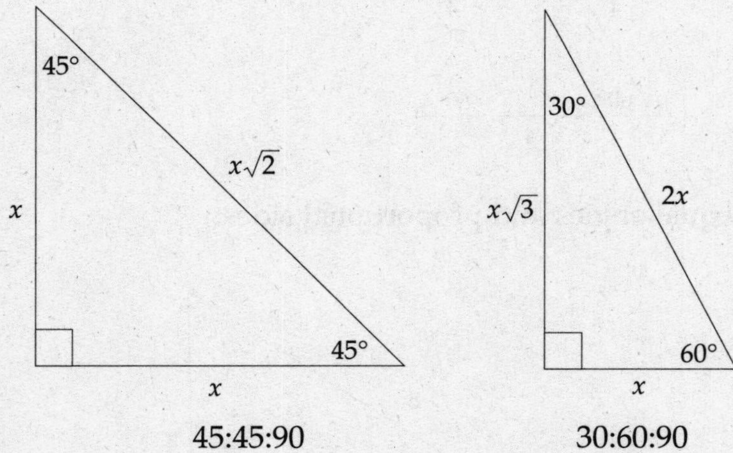

45:45:90 30:60:90

TIP: Memorize the shapes of these two triangles, and it will be easier to recognize them on the SAT. A 45:45:90 triangle is half of a square, and a 30:60:90 triangle is half of an equilateral triangle.

QUICK QUIZ #23

EASY

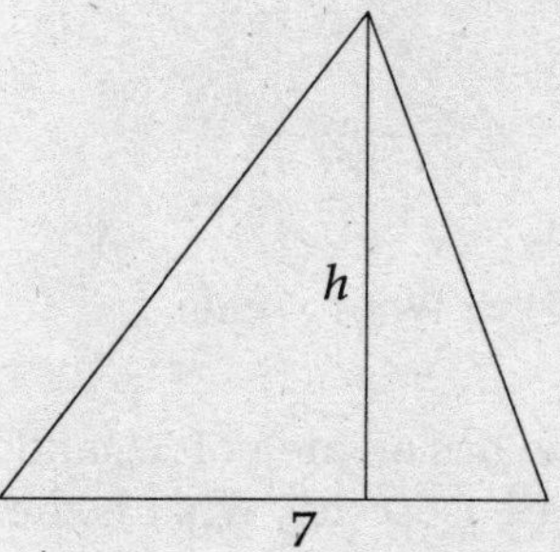

5 If the triangle above has an area of 21, then h equals

(A) 3 (B) 4 (C) 6 (D) 7 (E) 8

MEDIUM

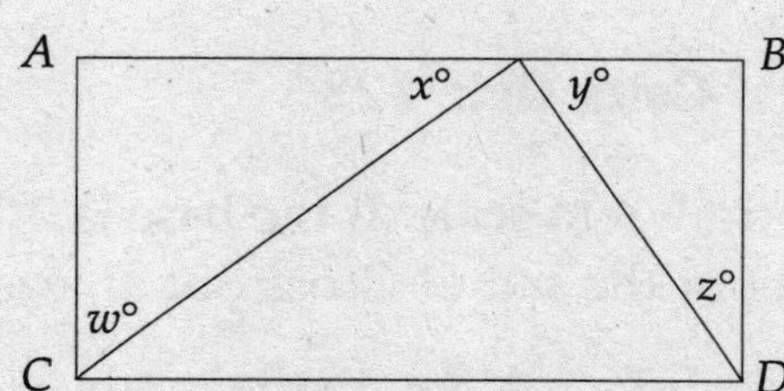

14 If ABCD is a rectangle, what is the value of $w + x + y + z$?

(A) 90 (B) 150 (C) 180 (D) 190 (E) 210

HARD

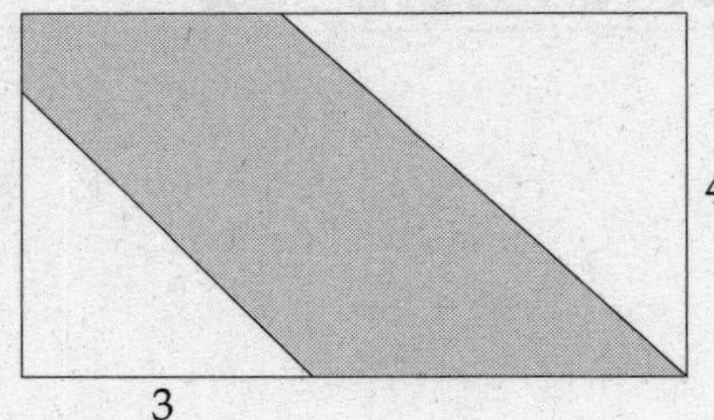

Note: Figure not drawn to scale.

22 If the rectangle above has an area of 32, and the unshaded triangles are isosceles, what is the perimeter of the shaded area?

(A) 16
(B) $10 + 7\sqrt{2}$
(C) $10 + 12\sqrt{2}$
(D) 32
(E) $70\sqrt{2}$

Answers and Explanations: Quick Quiz #23

5 C Estimate first—it's drawn to scale. If the base is 7, how long does the height look? About the same? Cross out at least A and E, and B if you're feeling confident. Now do the math: area = $\frac{1}{2}bh$, so $\frac{1}{2}(7h) = 42$, and $h = 6$. It would be easy to pick A if you weren't paying attention, because $7 \times 3 = 21$, and so it seems appealing. But only if you're aren't paying attention. And we know you are.

14 C If you picked A or E you didn't estimate. Very bad. See how the rectangle is cut up into 3 triangles? Each of those triangles has 180°. Both of the triangles with marked angles also have right angles, because they're corners of a rectangle.
So $\Delta ACE + \Delta EBD = 360°$. Subtract the 2 right angles, and you're left with 180°. Hot diggity dawg.

22 **B** First write in everything you know: if the area is 32, the length is 8. That means the base is 3 + 5 and the left side is 1 + 3. The triangles in opposing corners are both 45:45:90 triangles: the one on the base has a hypotenuse of $3\sqrt{2}$, and the one with sides of 4 has a hypotenuse of $4\sqrt{2}$. Add up all the sides of the shaded part, and you get $10 + 7\sqrt{2}$.

Here's how it should look:

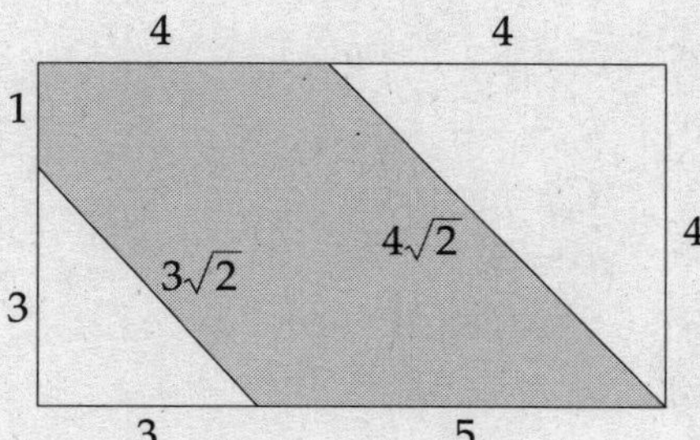

3. CIRCLES

Circles have 360°. Area = πr^2.

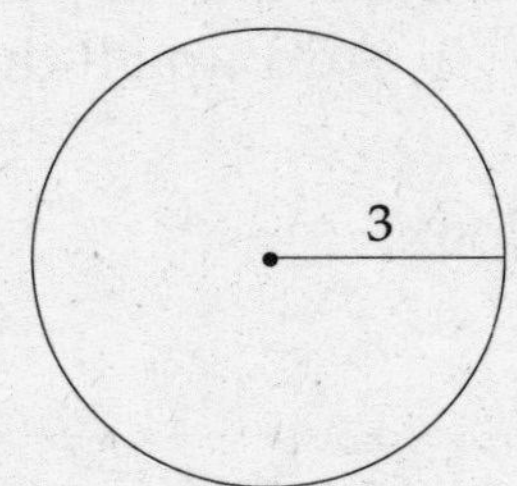

Circumference = $2\pi r$
$r = 3$
$C = 2\pi(3) = 6\pi$
$A = \pi(3)^2 = 9\pi$

TIP: For any pie slice of a circle, the central angle, arc, and area are in proportion to the whole circle:

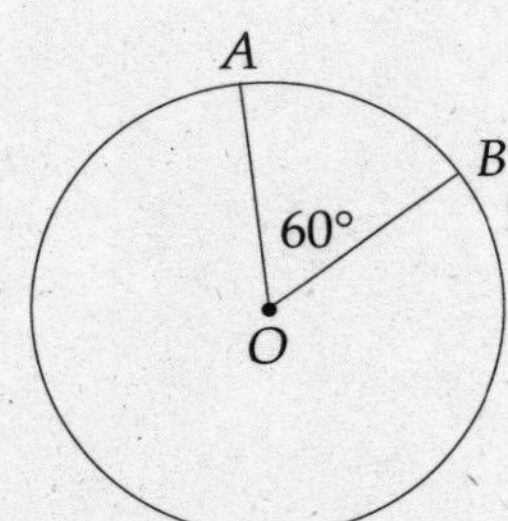

$\frac{60°}{360°} = \frac{1}{6}$, so arc AB is $\frac{1}{6}$ of the circumference and pie slice *AOB* is $\frac{1}{6}$ of the total area.

QUICK QUIZ #24

EASY

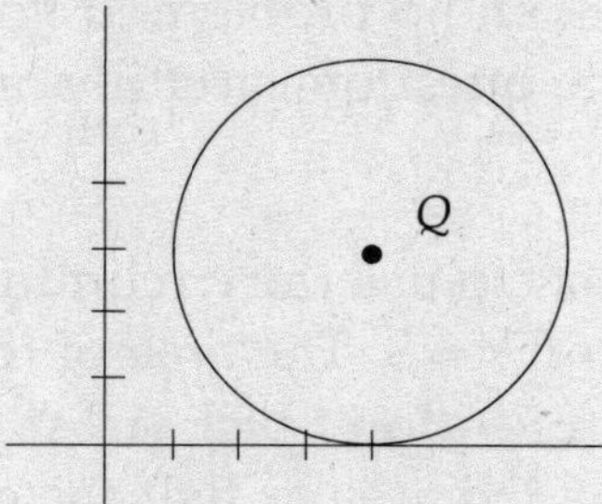

2π(4)

4 Center Q of the circle above has coordinates of (4, 3). What is the circumference of the circle?

(A) π (B) 2π (C) 6π (D) 8π (E) 9π

MEDIUM

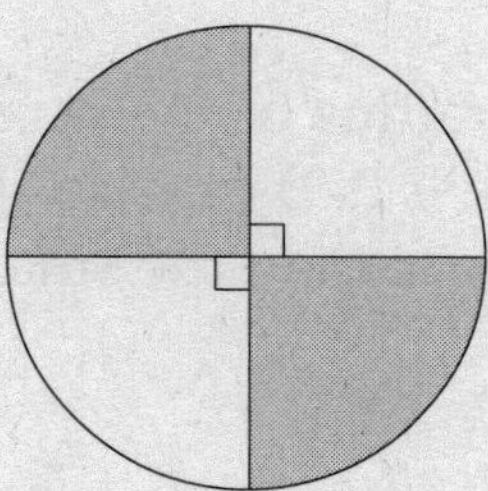

13 If the circumference of the circle above is 16π, what is the total area of the shaded regions?

π(4)²

(A) 64π (B) 32π (C) 12π (D) 8π (E) 4π

HARD

20 One circle has a radius of r, and another circle has a radius of $2r$. The area of the larger circle is how many times the area of the smaller circle?

(A) .5 (B) 1.5 (C) 2 (D) 3 (E) 4

Answers and Explanations: Quick Quiz #24

4 C The easiest way to solve this is simply to count the number of units in the radius, which is 3. Make sure you draw a radius on the diagram—if you draw it perpendicular to the *y*-axis you'll be able to count the units with no problem. If you picked E, you found the area. Read the question carefully and give 'em what they ask for.

13 B The circumference is 16π, so use the circumference formula to get the radius: $2\pi r = 16\pi$, and $r = 8$. The area of the whole circle is $\pi r^2 = \pi(8)^2 = 64\pi$. Hold on—don't pick A. At this point you could happily estimate the shaded area as half the circle and pick B. (Nothing else is close.) In fact, the shaded area is exactly half of the circle, because each marked angle is 90°, which makes each of those pie slices $\frac{90°}{360°}$ or $\frac{1}{4}$ of the circle. So two of them make up $\frac{1}{2}$ of the circle, or 32π. Trust what your eyes tell you.

20 E Plug in. If $r = 2$, then the radius of the small circle is 4π. The radius of the second circle is 2(2) or 4, so the area is 16π. The larger circle is 4 times bigger than the smaller circle. (Don't you just love to plug in?)

TIP: Notice how the hard question doesn't give you a picture or any real numbers to use. So draw the picture and make up your own numbers. Try to visualize the problem. Plugging in works just as well on geometry problems as it does on algebra problems.

A very common careless error on circle problems is getting the area and circumference mixed up. Memorize the formulas. We know, the formulas are given at the beginning of each section—but you'll waste time and confuse yourself if you have to keep flipping back to the front of the section in the middle of working out a problem.

4. QUADRILATERALS

All quadrilaterals have 360°.

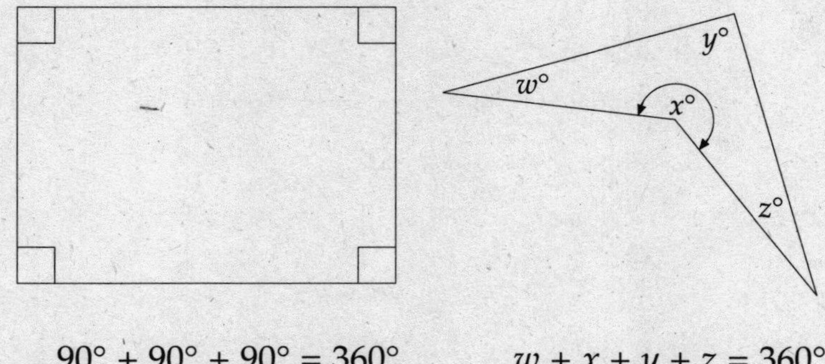

$90° + 90° + 90° = 360°$ $\qquad$ $w + x + y + z = 360°$

Perimeter: add up the sides.

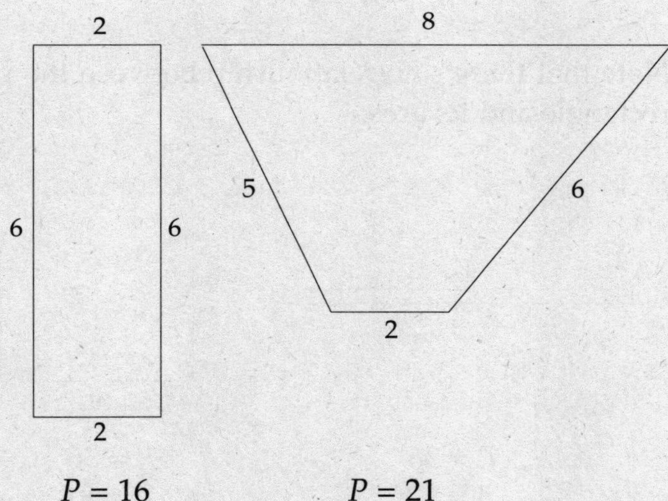

$P = 16$ $\qquad$ $P = 21$

RECTANGLES

All rectangles have four 90° angles and two pairs of parallel lines.

Area = $l \times w$

$A = lw$
$A = (9)(5)$
$A = 45$

TIP: Note that there's no relationship between the perimeter of a rectangle and its area:

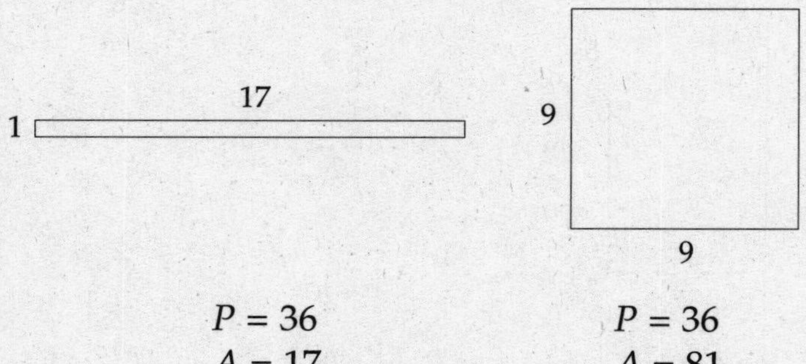

$P = 36$
$A = 17$

$P = 36$
$A = 81$

SQUARES

Squares have four 90° angles and two pairs of parallel lines, all of the same length.

Area = $l \times w$ or s^2

$A = s^2$

$A = 3^2$

$A = 9$

TIP: If you cut a square diagonally, you form two 45:45:90 triangles.

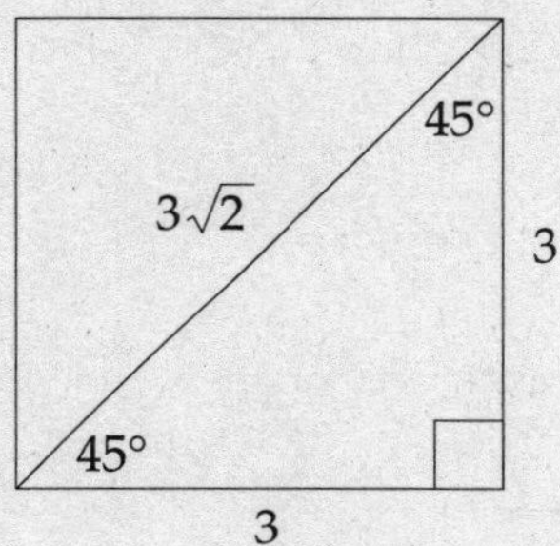

PARALLELOGRAMS

Parallelograms have two pairs of parallel lines, but no right angles.

Area = $b \times h$

$A = bh \quad A = (5)(4) \quad A = 20$

QUICK QUIZ #25

EASY

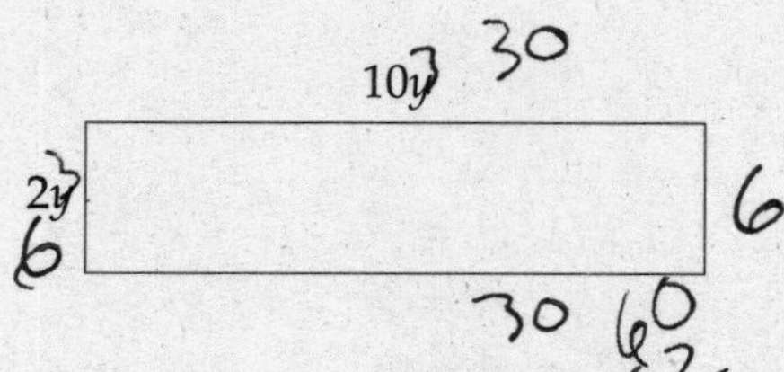

3 If $y = 3$, what is the perimeter of the figure above?

(A) 12 (B) 20 (C) 50 (D) 60 (E) 72

MEDIUM

15 The length of Rectangle A is $\frac{1}{3}$ the length of Rectangle B, and the width of A is twice the width of B. What is the ratio of the area of A to the area of B?

(A) $\frac{1}{3}$ (B) $\frac{2}{3}$ (C) 1 (D) $\frac{3}{2}$ (E) $\frac{3}{4}$

HARD

23 A hollow wooden cube has walls one inch thick. If the outer surface area of the cube is 150 square inches, then what is its inner surface area in square inches?

(A) 125 (B) 96 (C) 64 (D) 54 (E) 27

Answers and Explanations: Quick Quiz #25

6 **E** Figure out the dimensions of the rectangle if $y = 3$. That makes the length $10 \times 3 = 30$, and the width $2 \times 3 = 6$. Write those numbers on the diagram where they belong. To get the perimeter, add up all the sides. $30 + 30 + 6 + 6 = 72$.

15 **B** Let's plug in. If the length of A is $\frac{1}{3}$ the length of B, let's make the length of $B = 6$ and the length of $A = 2$. If the width of A is twice the width of B, let's make the width of $B = 4$ and the width of $A = 8$. Now the area of $A = 2 \times 8 = 16$, and the area of B is $6 \times 4 = 24$. $\frac{A}{B}$ is $\frac{16}{24}$, or $\frac{2}{3}$.

23 **D** Begin by drawing a cube so you see what's going on. Since a cube has 6 faces, its surface area is equal to the area of one face times 6. Let s represent the length of a face $150 = 6(s^2)$, so $s = 5$. Now take 1 inch off the sides of each face. Notice that you're really taking 2 inches from the length and the width, so the lengths of the inner faces equal 3. Then, $6(3^2) = 54$.

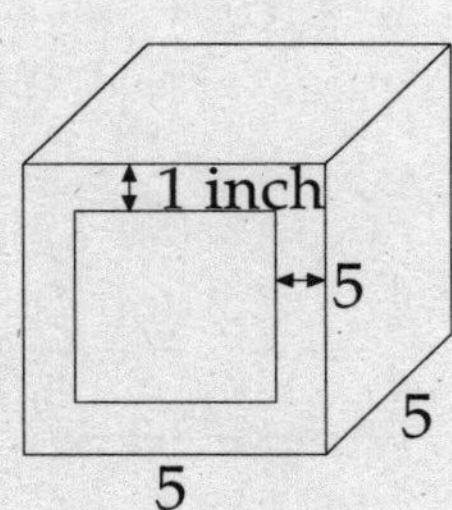

5. BOXES AND CANS

Forget spheres, cones, and other complicated 3-D nightmares. Most often, you will be asked only to deal with rectangular solids (boxes), cubes (square boxes), and possibly cylinders (cans).

No matter what the shape is, the volume equals the area × the third dimension (the depth or the height). Here are the formulas you need to know:

Rectangular Box
volume = $l \times w \times h$

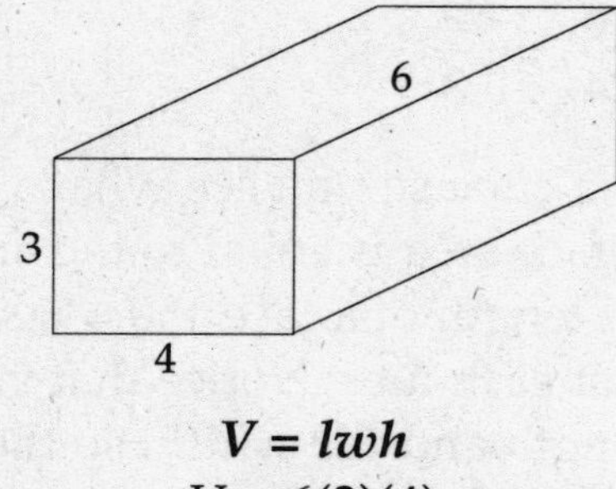

$$V = lwh$$
$$V = 6(3)(4)$$
$$V = 72$$

Cube
volume= s^3

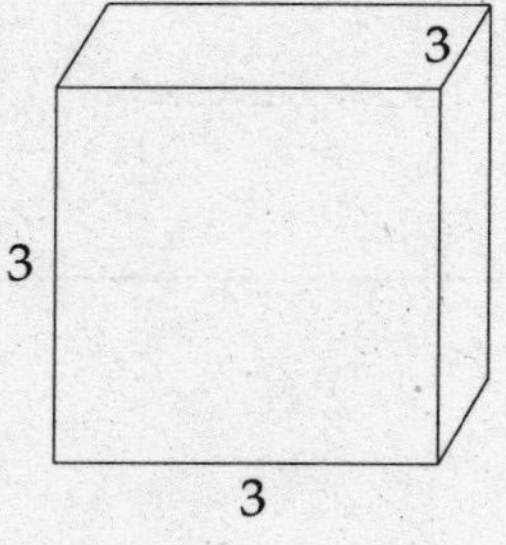

$$V = s^3$$
$$V = 3^3$$
$$V = 27$$

To find the diagonal of a box, draw in 2 right triangles: one on the end of the box, and the other cutting through the box. The second triangle will have the hypotenuse of the first triangle as its base, the length of the box as its height, and the diagonal of the box as its hypotenuse. Here's how it will look:

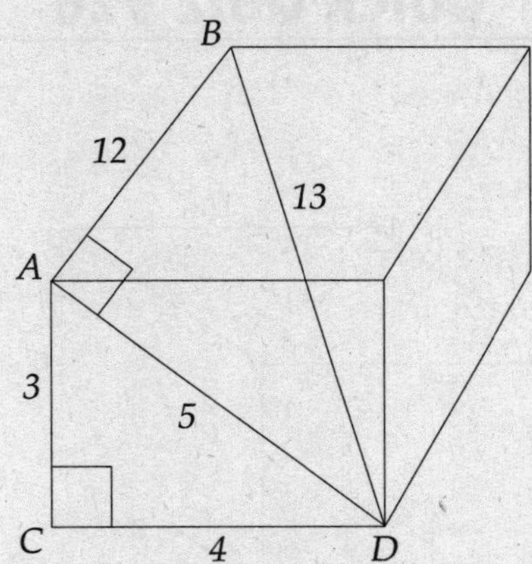

If AC is 3 and DC is 4, then AD is 5 (Pythagorean triple). If AD is 5 and AB is 12, then BD (the diagonal) is 13. (Another Pythagorean triple.) As you can probably guess, this only shows up on hard questions, and not that often. You can also estimate the length of the diagonal—it will be a little longer than the longest edge of the box.

Cylinder
volume= $\pi r^2 h$

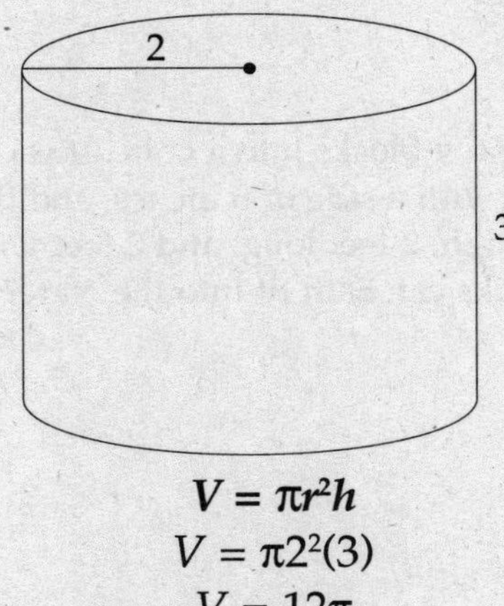

$$V = \pi r^2 h$$
$$V = \pi 2^2(3)$$
$$V = 12\pi$$

TIP: If the problem concerns a cone, pyramid, or any shape other than the ones described above, the necessary formula will be given in the question. If you aren't given a formula, you don't need one.

QUICK QUIZ #26

EASY

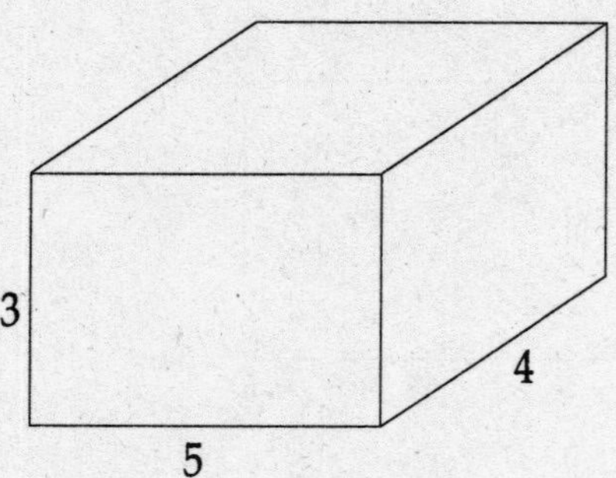

6 If the volumes of the two boxes above are equal, then h equals

(A) 1
(B) 2
(C) 4
(D) 5
(E) 20

MEDIUM

9 Sam is packing toy blocks into a crate. If each block is a cube with a side of 6 inches, and the crate is 1 foot high, 2 feet long, and 2 feet wide, how many blocks can Sam fit into the crate?

(A) 6
(B) 12
(C) 24
(D) 32
(E) 40

HARD

23 For an art project, Mark painted different sections of the interior and exterior of an 8-inch square box that had no top. First he painted one interior face blue and 2 exterior faces blue. Then he divided the 2 other exterior faces into 4 squares each, and painted $\frac{1}{4}$ of the squares blue. If Mark left the exterior bottom unpainted, and painted all other remaining faces green, what was the total area, in inches, of the green sections of the box?

(A) 128
(B) 224
(C) 288
(D) 352
(E) 512

Answers and Explanations: Quick Quiz #26

6 **A** The box on the left has volume = $3 \times 4 \times 5 = 60$. The box on the right is then $10 \times 6 \times h = 60$. So $h = 1$. Don't forget to estimate!

9 **D** First draw the crate. It should look like this:

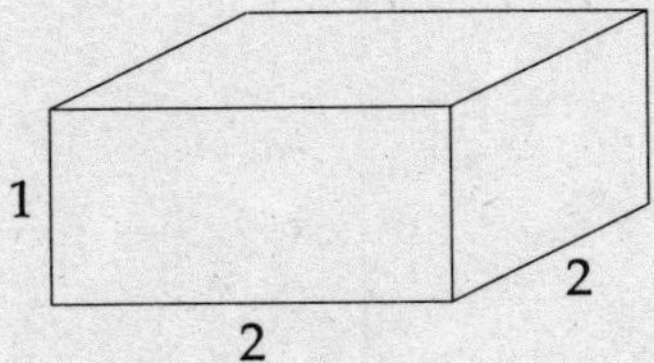

Now visualize putting blocks into the crate. If the blocks are 6 inches high, you'll be able to stack 2 rows in the crate since the crate is a foot high. Now mark off 6-inch intervals along the side of the crate. (You're dividing 2 feet, or 24 inches, by 6 inches.) You can fit 4 blocks along each side. Now multiply everything together and you get $2 \times 4 \times 4 = 32$ blocks.

23 D First draw yourself a cube, a nice-sized one, so you can visualize what's going on. You need to count up all the faces—that's 4 painted exterior faces and 4 painted interior faces, plus the interior bottom, for a total of 9 faces. If each face has a side of 8, then each has an area of 64. So all 9 faces have an area of 576. (Use your calculator.) Now you need to count up all the blue parts, subtract that from 576, and you'll be left with the green. Three faces are all blue (1 interior, 2 exterior). That's $3 \times 64 = 192$ blue so far. Now on your diagram, divide one face into 4 squares. If Mark did that on 2 faces, that's 8 little squares, each with an area of 16. $\frac{1}{4}$ of 8 is 2, so he painted only 2 little squares. Add $16 + 16$ to your 192. Now subtract 224 from 576, and finally, you get 352.

---

6. COORDINATE GEOMETRY

Remember how to plot points? The first number is x and the second is y.

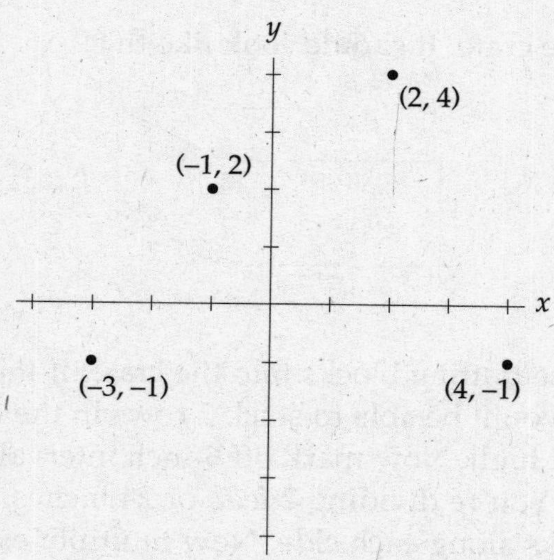

To find the length of a horizontal or vertical line, count the units:

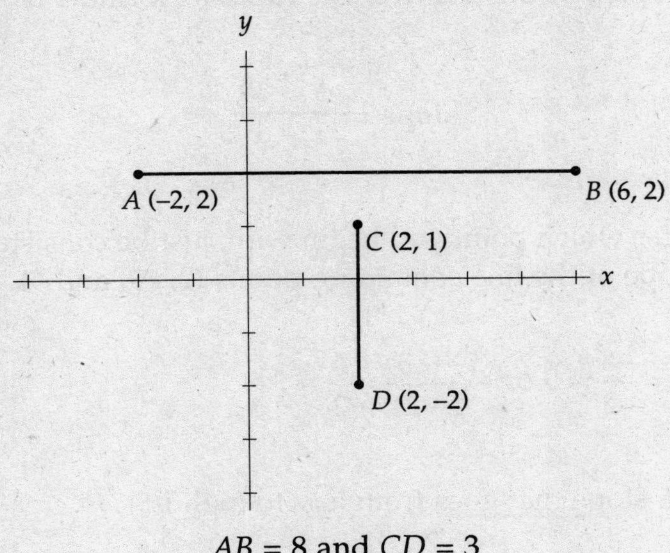

$AB = 8$ and $CD = 3$

To find the length of any other line, draw in a right triangle and use the Pythagorean Theorem:

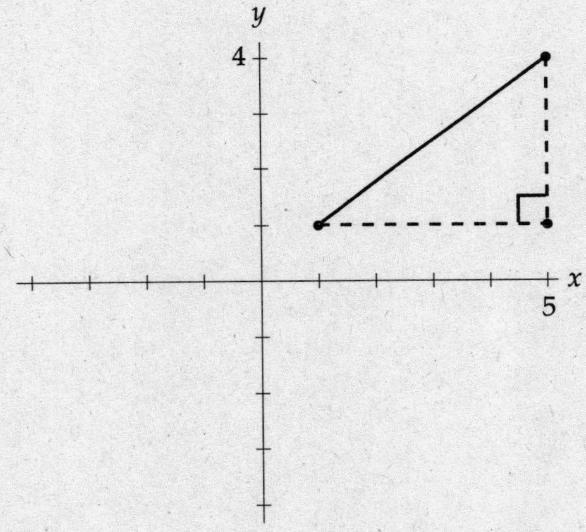

The triangle has legs of 3 and 4, so $3^2 + 4^2 = c^2$, and $c = 5$. (It's a Pythagorean triple again.)

To find the slope, put the rise over the run. The formula is

$$\textbf{slope} = \frac{y_1 - y_2}{x_1 - x_2}$$

It doesn't matter which point you begin with, just be consistent. What is the slope of the line containing points (2, –3) and (4, 3)?

$$\text{slope} = \frac{-3-3}{2-4} = \frac{-6}{-2} = 3 \text{ or } \frac{3-(-3)}{4-2} = \frac{6}{2} = 3$$

TIP: A slope that goes from low to high is +.

A slope that goes from high to low is –.

A slope that goes straight across is 0.

QUICK QUIZ #27

EASY

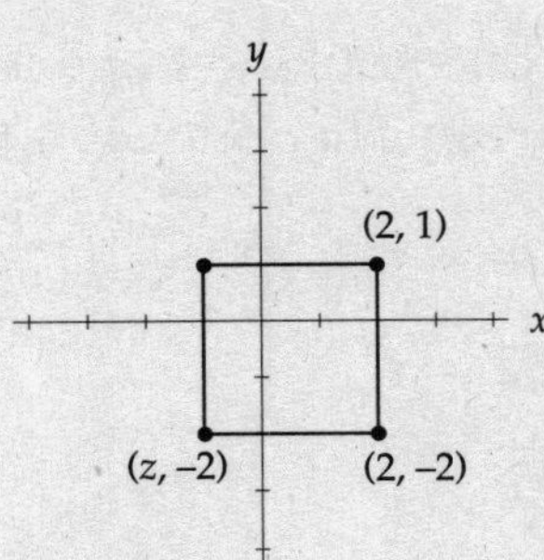

4 If the figure above is a square, what is the value of z?

(A) –2 (B) –1 (C) 1 (D) 2 (E) 4

MEDIUM

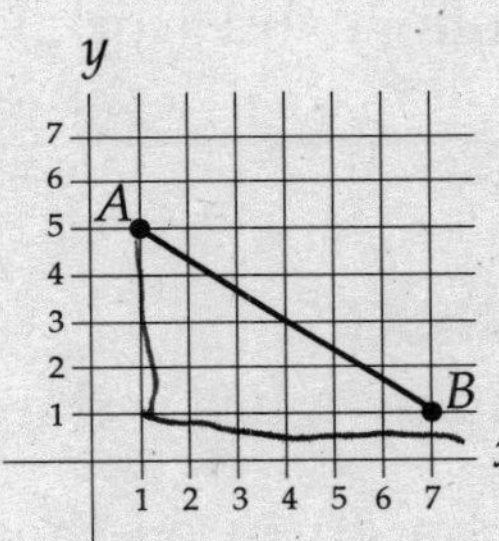

14 What is the length of AB?

(A) 4 (B) $2\sqrt{6}$ (C) 7 (D) $\sqrt{52}$ (E) $\sqrt{63}$

HARD

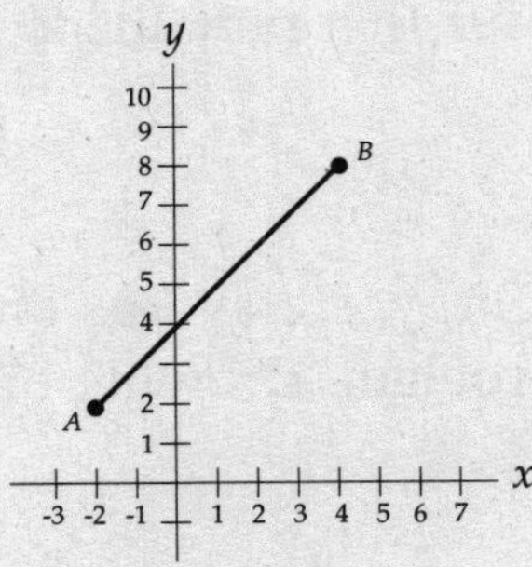

19 The coordinates for point A are (–2, 2) and the coordinates for point B are (4, 8). If line CD, not shown, is parallel to the line AB, what is the slope of line CD?

(A) –1 (B) 0 (C) 1 (D) 2 (E) 4

Answers and Explanations: Quick Quiz #27

4 **B** Just count the units. Remember that coordinates in the lower left quadrant will always both be negative.

14 **D** The units is each leg—you should get one leg = 4 and the other = 6. Now use the Pythagorean Theorem: $4^2 + 6^2 = c^2$

$$16 + 36 = c^2$$

$$52 = c^2$$

$$\sqrt{52} = c$$

19 **C** Write in the coordinates of A and B. $A = (-2, 2)$ and $B = (4, 8)$. So the slope of $AB = \frac{2-8}{-2-4} = \frac{-6}{-6} = 1$. If CD is parallel to AB, it has the same slope. (You could draw in a parallel line and recalculate the slope, but you'd be doing extra work.)

GEOMETRY: FINAL TIPS AND REMINDERS

- Always estimate first when the figure is drawn to scale.
- Always write the information given on the diagram, including any information you figure out along the way.
- If you don't know how to start, just look and see what shapes are involved. The solution to the problem will come through using the information we've gone over that pertains to that shape.

7. VISUAL PERCEPTION

To get this type of question right, you rely less on geometry formulas or rules than on visualizing the problem. Drawing on the diagram will help you do this.

For example:

What is the greatest number of regions that can be formed if 3 distinct lines intersect the circle?

(A) 4
(B) 5
(C) 6
(D) 7
(E) 8

Try drawing in the 3 lines. Count up how many regions you have. This is a hard question, so it won't be as simple as

The right way to draw the 3 lines is something like this:

That way you get 7 regions, which is the most you can get.

QUICK QUIZ # 28

EASY

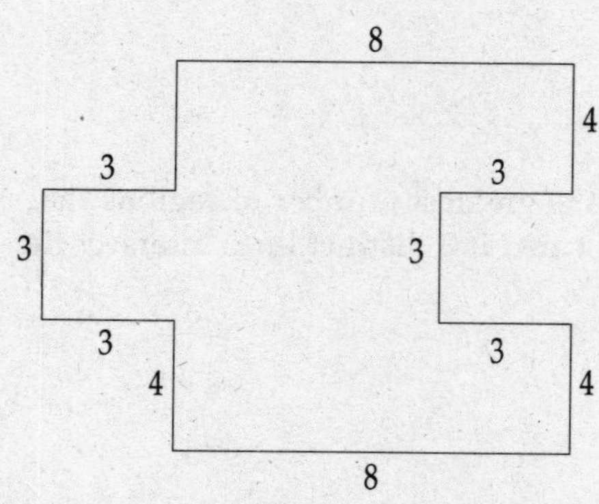

7 What is the area of the figure above, if all of the angles shown are right angles?

(A) 38
(B) 42
(C) 50
(D) 88
(E) 96

MEDIUM

A.

B.

C.

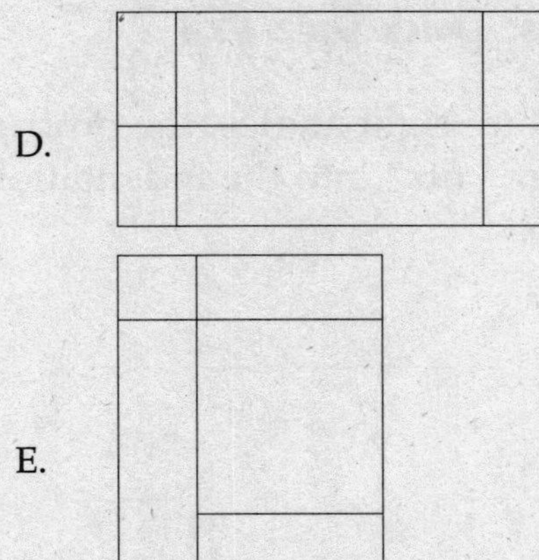

12 Which of the rectangles contains the greatest number of rectangles?

(A) A
(B) B
(C) C
(D) D
(E) E

HARD

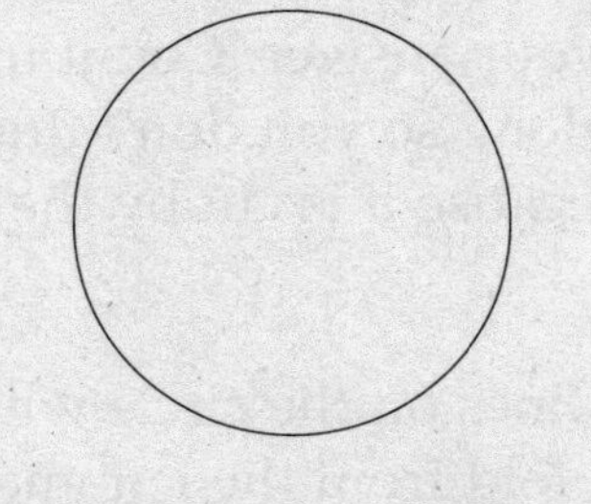

l_1

19 The circle above is folded from opposite sides so that the folds are perpendicular to l_1, and the folded edges of the circle meet. Then two more folds are made from opposite sides so that the folds are parallel to l_1. After all four folds are made, the circle becomes a

(A) cylinder
(B) rectangle
(C) circle
(D) triangle
(E) parallelogram

Answers and Explanations: Quick Quiz #28

7 **D** Since all the angles are right angles, the protruding piece in the left side of the figure "fits" into the indentation on the right, forming a rectangle.

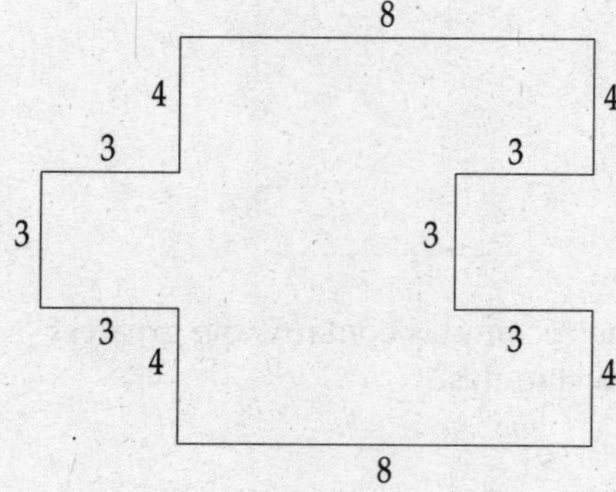

The area of the rectangle = length × width.

$$A = 8\ (4 + 3 + 4) = 8\ (11) = 88$$

12 **D** This kind of question can make you feel like you're losing your mind. One danger is spending too much time on it because it's hard to feel certain of your answer. Count up the rectangles, and mark them as you go along so you don't double-count. D has the most rectangles (16!) because it is cut by the most number of perpendicular lines.

19 **B** It helps to draw in the lines on the circle where you're going to fold it over. When you fold from the left and then the right, perpendicular to l^1, you get straight sides on the left and right and curves at top and bottom. Then when you fold down the top and the bottom, parallel to l^1, you end up with a rectangle. (Still unsure? Cut the circle out of the book with a pair of scissors and try folding it.)

CHAPTER 5

Special Question Formats

SPECIAL QUESTION FORMATS

1. QUANTITATIVE COMPARISON

The first 15 questions of one math section will be quantitative comparison questions. That's just a scary fancy name for this particular format. You'll see 2 columns, Column A and Column B, with something in each column. You decide whether Column A or B is bigger, if they're equal, or if you can't tell either way.

This section is full of sneaky traps, especially towards the end. Remember, on hard questions, to avoid picking answers that seem obvious, and try to estimate when you can. Avoid doing a lot of heavy calculating, especially if you can't do it on your calculator.

Directions: Pick A if Column A is always bigger.
Pick B if Column B is always bigger.
Pick C if they're always equal.
Pick D if you can't tell which is bigger.

Note: D usually means that one column is bigger only some of the time. D usually *doesn't* mean that you have no idea what the quantities could be.

EASY

	Column A	Column B
2	$6x - 6$	$6x + 6$
4	$\frac{1}{2}+\frac{2}{3}-\frac{1}{4}$	$\frac{2}{3}-\frac{1}{2}+\frac{1}{4}$

Solutions: For Question 2, it doesn't matter what $6x$ is. Since it's in both columns, it won't make either side bigger or smaller. But Column A has a –6 and Column B has a +6, so Column B is bigger. The answer is B. (Did that seem easy? Good.)

For Question 4, don't waste time figuring out the value of each side.

Instead, cross out $\frac{2}{3}$ since it's on both sides. Now compare what's left. In Column A you have $\frac{1}{2} - \frac{1}{4}$; in Column B you have $\frac{1}{4} - \frac{1}{2}$. In Column A you're subtracting a smaller fraction from a larger fraction; in B you're subtracting a larger fraction from a smaller one. The answer is A.

MEDIUM

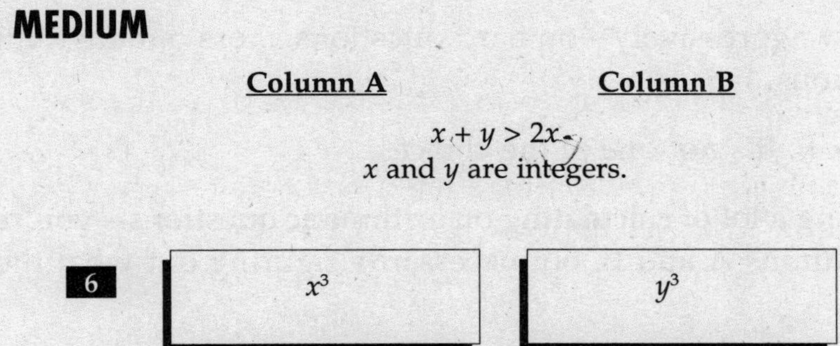

Solution: First manipulate the inequality to get $y > x$. Now which is bigger? Right—Column B is bigger. The answer is B.

HARD

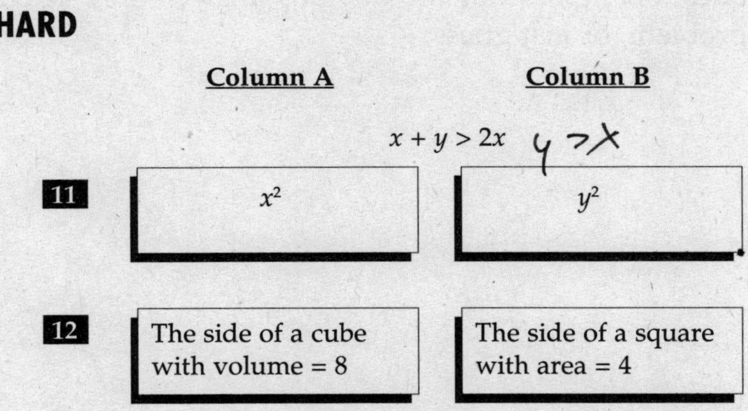

Solutions: Does Question 11 look familiar? We sure hope so—it's just like the medium question you just did, only this time x and y are squared, not cubed. So if $x = -2$ and $y = 1$, x^2 is bigger than y^2. If $x = 1$ and $y = 2$, then y^2 is bigger than x^2. Whenever you get an inconsistency like that, the answer is D. Bad guesses are B, because that's treating the question like a medium difficulty question, and C, for figuring that two squared things will be equal.

On Question 12, half the world is going to pick A because 8 is bigger than 4. Not you. If a cube has a volume of 8, $v = s^3$ and the side is then 2. In Column B, the side of the square with area of 4 is 2. They're equal, so the answer is C.

TIPS FOR QC HAPPINESS

- Write out A B C D next to each problem and cross them out as you go along.
- Guess very aggressively—on hard questions, cross out anything that looks obvious.
- Don't pick E. It's not one of the choices.
- Avoid doing a lot of calculating on arithmetic questions—you're *comparing* Columns A and B, not necessarily figuring out what they are exactly.
- For geometry questions with no picture, draw one yourself. On hard questions, there may be more than one way to draw the picture.
- Don't pick D if the question seems impossibly, ridiculously difficult.
- On hard questions, cross out the obvious answer first. Then try to solve the problem, or just guess.

QUICK QUIZ #29

	Column A	Column B
EASY		
1	$\frac{2}{3} - \frac{1}{2}$	$\frac{1}{2} - \frac{2}{3}$

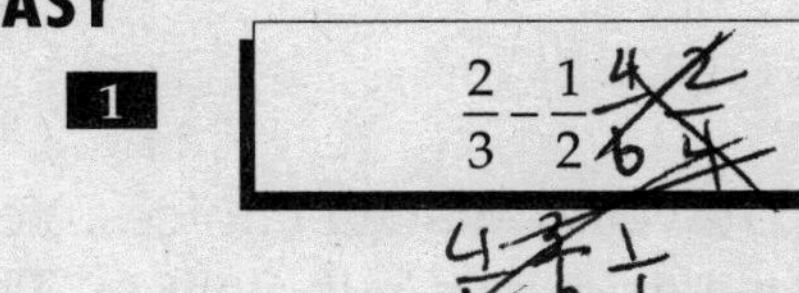

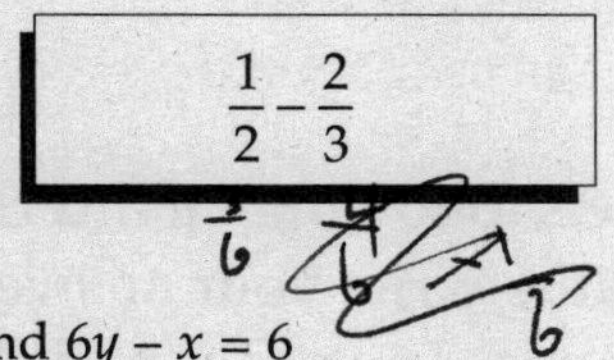

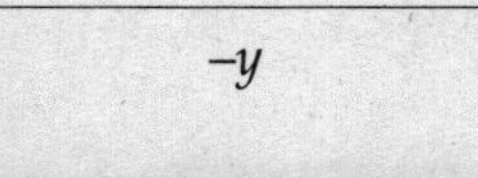

$2x - 5y = -6$ and $6y - x = 6$

	Column A	Column B
MEDIUM		
10	x	$-y$
HARD		
15	The area of Rectangle *A* with perimeter 36	The area of Rectangle *B* with perimeter 36

Answers and Explanations: Quick Quiz # 29

1 A Don't get overexcited and pick C by mistake. Column B is a negative number, since you're subtracting a bigger thing from a smaller thing.

10 C Stack 'em and add:

$$\begin{array}{r} 2x - 5y = -6 \\ \underline{+ (-x + 6y = 6)} \\ x + y = 0 \\ x = -y \end{array}$$

(When you get to $x + y = 0$, you can plug in.)

15 **D** No relationship between area and perimeter of a rectangle! The dimensions of a rectangle with perimeter 36 could be 9 and 9; 12 and 6; 10 and 8; etc., etc., etc. Those dimensions give you different areas, and we have no idea what the dimensions of either A or B are.

---

2. GRID-INS

The ten questions following quant comp have no answer choices. You must solve the question, write your answer on a grid, and bubble it in. This isn't as bad as it sounds. The order of difficulty applies, so the first 3 questions (16–18) are easy, the middle 4 questions (19–22) are medium, and the final 3 questions (23–25) are hard. Take your time on the easy and medium quesitons, as always.

TIPS FOR GRID-IN HAPPINESS

- Don't bother to reduce fractions: $\frac{3}{6}$ is as good as $\frac{1}{2}$.
- Don't round off decimals. If your answer has more than four digits, just start to the left of the decimal point and fit in as many as you can.
- Don't grid in mixed fractions. Either convert to one fraction or a decimal. (Use 4.25 or $\frac{17}{4}$, not $4\frac{1}{4}$.)
- If the question asks for "one possible value," any answer that works is OK.
- Forget about: negatives, variables, π, $, and %. You can't grid 'em.
- You can still plug in if the question has an implied variable (like the Mr. Heftwhistle problem on page 8).

QUICK QUIZ #30

EASY

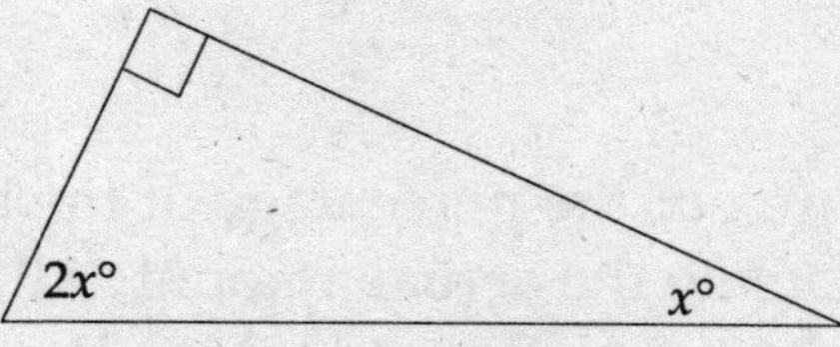

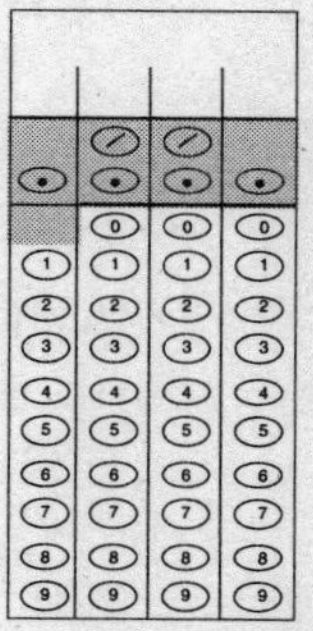

17 What is the value of x?

MEDIUM

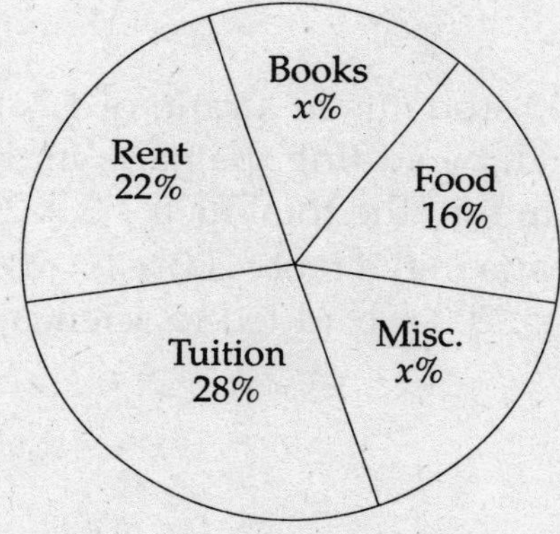

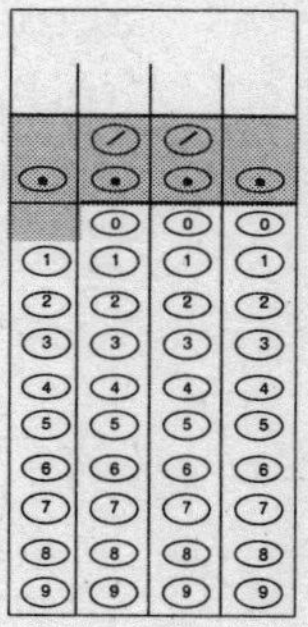

20 The chart above shows Orwell's projected expenditures for his freshman year at River State University. If he plans to spend a total of $10,000 for the year, how many dollars will Orwell spend on books?

HARD

24 At a certain university, the ratio of science majors to art majors was 3:1. If 600 other students changed their majors to science and art in a ratio of 1:2, the new ratio of science to art majors is 4:3. What was the original number of art majors?

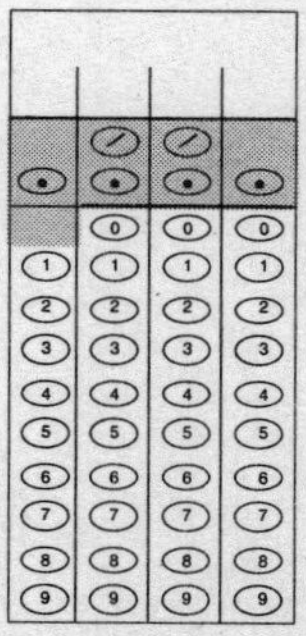

Answers and Explanations: Quick Quiz #30

17 **30**

Since this is a right triangle, the other 2 angles add up to 90. So $3x = 90$ and $x = 30$.

20 **1700**

Two steps: first figure out the percentage of the budget spent on books, and then calculate the actual amount. All the pie slices add up to 100%, so $28 + 16 + 22 + 2x = 100$. $2x = 34$ and $x = 17\%$. Take 17% of 10,000, which is 1700.

24 **200**

Write the ratio out so you can see what you're doing:

S:A

3:1

If we're adding in 600 students in a ratio of 1:2, figure out the actual numbers of each by adding the parts in the ratio ($1 + 2 = 3$) and dividing that sum into the total ($600 \div 3 = 200$). Now multiply that quotient by the parts in the ratio: $200 \times 1 = 200$ and $200 \times 2 = 400$. So out of 600 students, 200 are added to science and 400 are added to art:

```
   S:A
   3:1
+ 200:400
   4:3
```

The easiest thing to do next is to plug in numbers for the original ratio of 3:1. Since we're dealing in hundreds, let's try 300:100. With the new students added in, we get 500:500, or 1:1. We need to try something bigger, like 600:200. (Still maintaining our original 3:1 ratio.) Add in the new students, and you get 800:600, or 4:3. So the number of original art majors was 200.

CHAPTER

Problem Sets

PROBLEM SETS

The following groups of questions were designed for quick, concentrated study. The quant comp questions come in sets of 15, just like on the SAT, and the rest come in groups of ten (3 easy, 4 medium, 3 hard). Answers and explanations follow immediately. The idea is for you to check your answers right after working the problems, so that you can learn from your mistakes before you continue.

Don't simply count up how many you got wrong and then breeze on to the next thing—take a careful look at *how* you got the question wrong. Did you use the wrong strategy? Not remember the necessary basic math? Make a goofy computation error? Write an equation and backsolve at the same time?

You need to know the cause of your mistakes before you can stop making them.

We put together sets of plugging in, backsolving, and estimating questions to help you learn how to recognize those types of questions when they come up—so pay attention to the look and feel of them. For these strategy questions, we put the answers and explanations right after each level of difficulty, so you can be certain of mastering the strategy before going on to more difficult questions. When you get to Problem Set 6, the answers to all the questions in the set will follow the set immediately.

One last thing—the question numbers correspond exactly to the difficulty level in both the 10-question section, the quant comp problems, and the grid-in problems. For the 25-question multiple choice section, the easy questions are #1-8, the mediums are #9-17, and the hard ones are #18-25. We point this out because you must always be aware of the difficulty level of the question you're working on.

Tallyho!

PROBLEM SET 1: PLUGGING IN

EASY

1 Sinéad has 4 more than three times the number of hats that Maria has. If Maria has x hats, then in terms of x, how many hats does Sinead have?

(A) $3x + 4$
(B) $3(x + 4)$
(C) $4(x + 3)$
(D) $4(3x)$
(E) $7x$

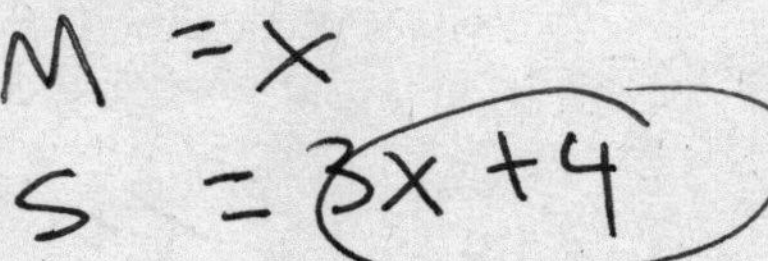

2 When 6 is subtracted from $10p$, the result is t. Which of the following equations represents the statement above?

(A) $t = 6(p - 10)$
(B) $t = 6p - 10$
(C) $t = 10(6 - p)$
(D) $10p - 6 = t$
(E) $10 - 6p = t$

3 Sally scored a total of $4b + 12$ points in a certain basketball game. She scored the same number of points in each of the game's 4 periods. In terms of b, how many points did she score in each period?

(A) $b - 8$
(B) $b + 3$
(C) $b + 12$
(D) $4b + 3$
(E) $16b + 48$

MEDIUM

4 If t is a prime number, and x is a factor of 12, then $\frac{t}{x}$ could be all of the following EXCEPT

(A) $\frac{1}{12}$ (B) $\frac{1}{4}$ (C) $\frac{1}{2}$ (D) 1 (E) 2

5 Roseanne is 6 years younger than Tom will be in 2 years. Roseanne is now x years old. In terms of x, how old was Tom 3 years ago?

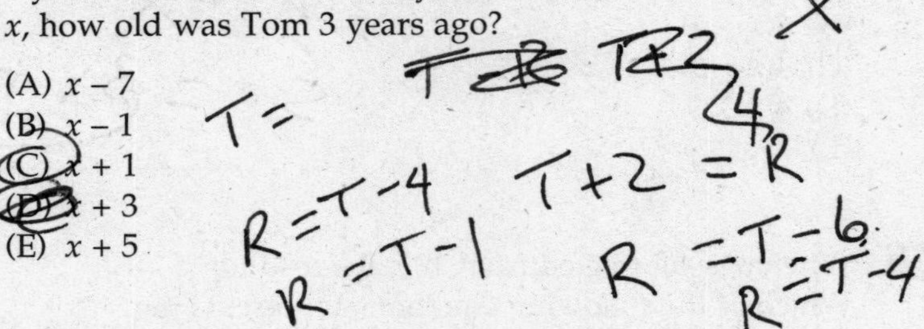

(A) $x - 7$
(B) $x - 1$
(C) $x + 1$
(D) $x + 3$
(E) $x + 5$

6 A phone company charges 10 cents per minute for the first 3 minutes of a call, and $10 - c$ cents for each minute thereafter. What is the cost, in cents, of a 10-minute phone call?

(A) $200 - 20c$
(B) $100c + 70$
(C) $30 + 7c$
(D) $100 - 7c$
(E) $100 - 70c$

7 If $0 < pt < 1$, and p is a negative integer, which of the following must be less than -1?

(A) p (B) $p - t$ (C) $t + p$ (D) $2t$ (E) $t \times \frac{1}{2}$

HARD

8 If x and y are positive integers, and $\sqrt{x} = y + 3$, then $y^2 =$

(A) $x - 9$
(B) $x + 9$
(C) $x^2 - 9$
(D) $x - 6\sqrt{x} + 9$
(E) $x^2 - 6\sqrt{x} + 9$

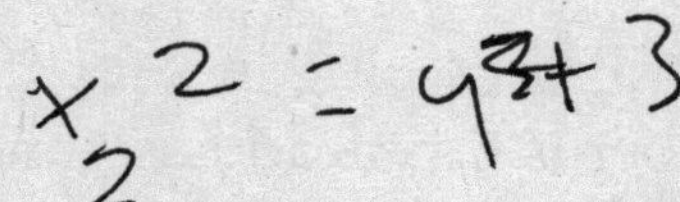

9 If cupcakes are on sale at 8 for c cents, and gingerbread squares are on sale at 6 for g cents, what is the cost, in cents, of 2 cupcakes and 1 gingerbread square?

(A) $8c + 3g$

(B) $\frac{cg}{3}$

(C) $\frac{8c+6g}{3}$

(D) $\frac{8c+3g}{14}$

(E) $\frac{3c+2g}{12}$

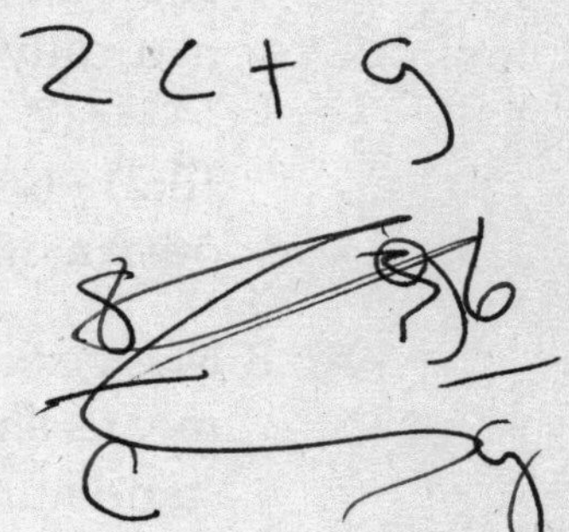

10 If the side of a square is $x + 1$, then the diagonal of the square is

(A) $x^2 + 1$
(B) $2x + 2$
(C) $x\sqrt{2} + \sqrt{2}$
(D) $x^2 + 2$
(E) $\sqrt{2x} + \sqrt{2}$

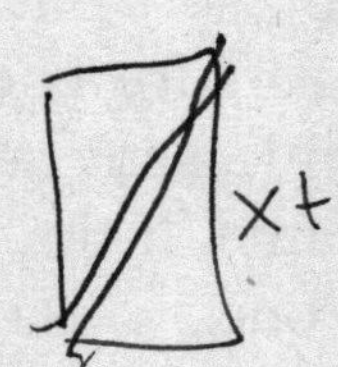

Answers and Explanations: Problem Set 1

EASY

1 **A** Forget the algebra. Let's plug in 2 for x, so Maria has 2 hats. Triple that number is 6. Sinead has 4 more than triple, so Sinead has $4 + 6 = 10$. You should put a circle around 10, so you can remember it's the answer to the question, the magic number. Now plug 2 into the answer choices. A gives us $3(2) + 4 = 10$, which is just what we're looking for.

2 **D** Plug in 2 for p. $2\times 10 = 20$; $20 - 6 = 14$. So $t = 14$. (Since this is an equation, when you pick one number, the other number is automatically produced by the equation.) If $p = 2$ and $t = 14$, A is $14 = 6(2 - 10)$. Does $14 = 12 - 60$? Not on this planet. D is $10(2) - 6 = 14$, or $20 - 6 = 14$. The equation works, so that's our answer.

3 **B** Maybe Sally should try out for the NBA. Let's make $b = 2$. That means she scored $4(2) + 12 = 20$ points total. If she scored the same number of points in each of the four periods, we have to divide the total by 4, so she scored $20 \div 4 = 5$ points per period. Put a circle around 5. Now on to the answer choices. A is $2 - 8$. B is $2 + 3 = 5$, which is our magic number.

TIP: Notice how we keep plugging in 2? That's because we're trying to make things as easy as possible. To get these questions right you didn't *have* to pick 2; on some questions, 2 might not work out so well. You can pick whatever you want. Just make sure your number doesn't require you to make ugly, unpleasant calculations. *Avoiding hard work* is the name of the game. If the number you pick turns bad on you, pick another one.

MEDIUM

4 **A** This question is tricky to spot as a plugging-in question because the answer choices don't have variables. Just remember that most of the time you can make things easier by plugging in numbers for variables in the question. Here we go: make a short list of possibilities for t, starting with the first prime number. Then do the same for x, listing the factors of 12 in pairs.

$t = 2, 3, 5, 7$ $\qquad\qquad$ $x = 1, 12, 2, 6, 3, 4$

The question asks us for $\frac{t}{x}$, which we can make by putting any number in our t column over any number from our x column. B is $\frac{3}{12}$. C is $\frac{3}{6}$. D is $\frac{2}{2}$. E is $\frac{2}{1}$. No matter what we do, we can't make $\frac{1}{12}$, so A is our answer. If you didn't remember that 1 is not prime, you were probably banging your head against a wall. When that happens, go on to the next question.

5 **C** Let $x = 10$, so Roseanne is now 10 years old. That's 6 years younger than 16, so Tom must be 16 in 2 years, which makes him 14 now. The question asks for Tom's age 3 years ago; if he's 14 now, 3 years ago he was 11. Circle 11. In the answer choices, plug 10 in for x. A is $10 - 3$. Nope. B is $10 - 1$. Nope. C is $10 + 1$. Yeah!

6 **D** Let $c = 8$. The first 3 minutes of the call would be 3(10) or 30 cents. The remaining minutes would be charged at $10 - 8$ cents, or 2 cents a minute. There are 7 minutes remaining, so $2 \times 7 = 14$. The total cost is $30 + 14 = 44$ cents. On to the answer choices: A and B are way too big. C is $30 + 7(8) = 86$. D is $100 - 56 = 44$. Bingo.

7 C First take a good look at $0 < pt < 1$. We know that pt is a positive fraction. If p is a negative integer, then t must be a negative fraction. Now we're ready to plug in. (Or you can just try numbers until you find some that satisfy the inequality.) Let $p = -1$ and $t = -\frac{1}{2}$. Try them in the answer choices, crossing out any answer that's –1 or higher. A is –1, cross it out. B is $-\frac{1}{2}$, cross it out. C is $-1\frac{1}{2}$, leave it in. D is –1, cross it out. E is –1, cross it out. The trouble with *must be* questions is that you can only *eliminate* answers by plugging in, you can't simply choose the first answer that works. That's because the answer may work with certain numbers but not with others—and you're looking for an answer that *must be true*, no matter what numbers you pick. These questions can be time-consuming, so if you're running low on time, you may want to skip them.

HARD

8 D Let $x = 25$. That makes $y = 2$. The question asks for y^2, and $2^2 = 4$. Circle it. Now try the answer choices. A is $25 - 9$. B is $25 + 9$. C is huge. D is $25 - 30 + 9 = 4$. Not so bad, huh?

9 E Let $c = 16$ and $g = 12$. That means the cupcakes and the gingerbread squares sell for 2 cents apiece. One gingerbread square and two cupcakes will cost 6 cents. Circle 6. On to the answer choices, plugging in 16 for c and 12 for g. E gives you $\frac{3(16)+2(12)}{12} = 6$. Use your calculator for that last part. Get it right? Then go to a bakery and celebrate.

10 C Draw yourself a little square and label the sides $x + 1$. Draw in a diagonal. Let $x = 2$. The side of the square is then 3, and the diagonal is $3\sqrt{2}$. (The diagonal is the hypotenuse of a 45-45-90 triangle.) Plug 2 into the answer choices. C gives you $2\sqrt{2} + \sqrt{2} = 3\sqrt{2}$.

PROBLEM SET 2: MORE PLUGGING IN

EASY

1 Jim and Pam bought x quarts of ice cream for a party. If 10 people attended the party, including Jim and Pam, and if each person ate the same amount of ice cream, which of the following represents the amount of ice cream, in quarts, eaten by each person at the party?

(A) $10x$ (B) $5x$ (C) x (D) $\frac{x}{5}$ (E) $\frac{x}{10}$

2 If x and y are integers and $\frac{x}{y} = 1$, then $x + y$ must be

(A) positive
(B) negative
(C) odd
(D) even
(E) greater than 1

3 If $3x - y = 12$, then $\frac{y}{3} =$

(A) $x - 3$
(B) $x - 4$
(C) $3x - 4$
(D) $9x - 12$
(E) $3x + 4$

MEDIUM

4 When x is divided by 3, the remainder is z. In terms of z, which of the following could be equal to x?

(A) $z - 3$
(B) $3 - z$
(C) $3z$
(D) $6 + z$
(E) $9 + 2z$

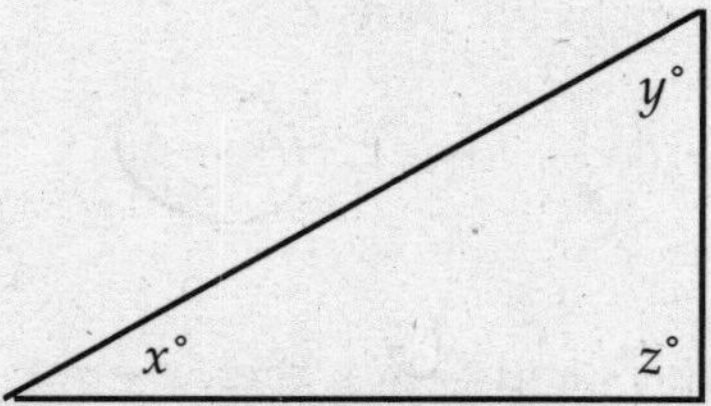

5 In the figure above, $2x = y$. In terms of x, $z =$

(A) $180 + 2x$
(B) $180 + x$
(C) $180 - x$
(D) $180 - 3x$
(E) $180 - 4x$

6 If w, x, y, and z are consecutive positive integers, and $w > x > y > z$, which of the following CANNOT be true?

(A) $x + z = w$
(B) $y + z = x$
(C) $x - y = z$
(D) $w - x = y$
(E) $w - z = y$

7 The volume of a certain rectangular solid is $12x$. If the dimensions of the solid are the integers x, y, and z, what is the greatest possible value of z?

(A) 36
(B) 24
(C) 12
(D) 6
(E) 4

HARD

8 If $x^3 < 0 < xy^2z$, which of the following must be true?

I. xyz is positive
II. $x^2y^2z^3$ is positive
III. $x^3y^2z^3$ is positive

(A) I only
(B) III only
(C) I and II only
(D) II and III only
(E) I, II, and III

9 When a is divided by 7, the remainder is 4. When b is divided by 3, the remainder is 2. If $0 < a < 24$ and $2 < b < 8$, which of the following could have a remainder of 0 when divided by 8?

(A) $\frac{a}{b}$ (B) $\frac{b}{a}$ (C) $a - b$ (D) $a + b$ (E) ab

10 If $3x$, $\frac{3}{x}$, and $\frac{15}{x}$ are integers, which of the following must also be an integer?

I. $\frac{x}{3}$

II. x

III. $6x$

(A) I only
(B) II only
(C) III only
(D) I and III only
(E) II and III only

Answers and Explanations: Problem Set 2

EASY

1 E Plug in 20 for x. If 10 people eat 20 quarts, and they all eat the same amount, then each person eats 2 quarts. (The pigs!) Put a circle around 2. Go to the answers and remember that $x = 20$.

$A = 10 \times 10 = 100$. Nope. $E = \frac{20}{10} = 2$. Yep.

2 D Plug in 2 for x and 2 for y. That satisfies the equation $\frac{x}{y} = 1$, and makes $x + y = 4$. Eliminate B and C. How about $-2 = x$ and $-2 = y$? That makes $x + y = -4$. Eliminate A and E.

3 B Plug in 5 for x, which makes $y = 3$. So $\frac{y}{3} = \frac{3}{3} = 1$. Circle 1. On to the answers, and plug in $x = 5$. $A = 5 - 3 = 2$, no good. $B = 5 - 4 = 1$. There you go.

TIP: Why do we keep saying "Circle it" in the explanations? Because that's the arithmetic answer to the question. All that's left to do is plug in for the variables in the answer choices, and look for your circled number. We tell you to circle that number so it won't get lost in the shuffle, and you can keep track of what you're doing.

Notice how sometimes, as in Question 2, you may have to plug in more than one set of numbers. That doesn't mean you're doing anything wrong, it's just the nature of the question—and it tends to happen on *must be* questions.

Also remember that you are trying to find numbers to plug in that make getting an answer to the question easy—so in Question 3, if we'd plugged in $x = 2$, that would've made y negative. Who wants to deal with negatives if they don't have to? If some kind of nastiness happens, bail out and *pick new numbers*.

MEDIUM

4 **D** Let $x = 7$ so $z = 1$. Try the answers. D is $6 + z$, or $6 + 1 = 7$. No sweat.

5 **D** Plug in 10 for x, which makes $y = 20$. Remember that a triangle has 180°, so the third angle, z, must equal $180 - 30 = 150$. Circle 150. Try the answers, with $x = 10$. D gives us $180 - 30 = 150$.

6 **D** First put the inequalities in order: $w > x > y > z$. Plug in consecutive positive numbers for the variables: $w = 5$, $x = 4$, $y = 3$, $z = 2$. Now try the answers.

A: $4 + 2 = 5$. No, so leave it in.

B: $3 + 2 = 4$. No, so leave it in.

C: $4 - 3 = 2$. No, so leave it in.

D: $5 - 4 = 3$. No, so leave it in.

E: $5 - 2 = 3$. Yes, so cross it out.

Whew. We didn't have very good luck. Let's try a new set of numbers. $w = 4$, $x = 3$, $y = 2$, and $z = 1$. This time A, B, and C all work, so we get rid of 'em—that leaves D as our answer. Watch out when the question says CANNOT; it's all too easy to get mixed up and start thinking in the wrong direction. Look for answers that *work*, and cross them out, rather than looking for the answer that *doesn't* work.

7 **C** First draw yourself a pretty little picture. (Think shoebox.) Plug in 2 for x. The formula for volume of a rectangular solid is length × width × height—in this case, xyz. Our volume is $12x = 12(2) = 24$. Let's come up with 3 different numbers—2 is one of them—that give us 24 when multiplied together.

A chart is never a bad idea. It keeps you organized.

x	y	z
2	1	12

Since y is as low as possible, z is as big as possible. Go with it. If you're not convinced, try other combinations—but don't forget, the question asks for the greatest possible value of z.

HARD

8 **B** Since we've got a lot of exponents, we want to plug in the smallest numbers we can. Positive/negative is what this quesion is about. Let's take it part by part. If $x^3 < 0$, we know x must be negative: let $x = -1$. If $-1(y^2z)$ is positive, let $z = -2$. Beware! We don't know whether y is positive or negative, so we can't plug anything in for it. Let's look at I, II, and III, and remember we're looking for something that *must be* true:

I. $xyz = -1(y)(-2)$. Maybe it's positive and maybe it isn't. It depends on the sign of y. And we don't know the sign of y.

II. $x^2y^2z = 1(y^2)(-2)$. Sorry. It's negative.

III. $x^3y^2z^3 = 1(y^2)(-2^3)$. Well, well, well. Since y is squared, it has to be positive. And the product of the other two negatives equals positive. It works.

9 **D** Plug in 11 for a and 5 for b. Those 2 choices satisfy all the conditions of the problem. Check the answers: A and B are fractions, forget about 'em. C is $11 - 5 = 6$, which isn't divisible by 8. D gives you $11 + 5 = 16$. If we divide 16 by 8, we get a quotient of 2 and a remainder of 0. End of story.

10 **C** How about plugging in 3 for x? Try the answers—we're looking for integers, so if the answer isn't an integer, we can cross it out.

I. $\frac{x}{3} = \frac{3}{3} = 1$ OK so far.

II. $x = 3$ OK so far.

III. $6x = 6 \bullet 3 = 18.$ OK so far.

At this point, your average test-taker figures the question is pretty easy and picks E. Not you, my friend. *This is a hard question.* You must go an extra step. Plug in a new number. Since the question concerns integers, what if we plug in something that isn't an integer? Like $x = \frac{1}{3}$?

I. $\dfrac{\frac{1}{3}}{3} = \dfrac{1}{9}$. That's no integer. Cross it out.

II. $\dfrac{1}{3}$. No good either.

III. $6\left(\dfrac{1}{3}\right) = 2$. OK.

There's more. What if $x = 15$? That gets rid of I, and you're left with III only.

PROBLEM SET 3: BACKSOLVING

EASY

1 If x is a positive integer and $x + 12 = x^2$, what is the value of x?

(A) 2 (B) 4 (C) 6 (D) 8 (E) 12

2 If twice the sum of three consecutive numbers is 12, and the two lowest numbers add up to 3, what is the highest number?

(A) 2 (B) 3 (C) 6 (D) 9 (E) 12

3 If $2^x = 8^{(x-4)}$, then $x =$

(A) 4 (B) 6 (C) 8 (D) 9 (E) 64

MEDIUM

4 If Jane bought 3 shirts on sale for the same price, she would have 2 dollars left over. If instead she bought 10 equally-priced pairs of socks, she would have 7 dollars left over. If the prices of both shirts and socks are integers, which of the following, in dollars, could be the amount Jane has to spend?

(A) 28 (B) 32 (C) 47 (D) 57 (E) 60

5 During a vacation together, Bob spent twice as much as Josh, who spent four times as much as Ralph. If Bob and Ralph together spent \$180, how much did Josh spend?

(A) \$20
(B) \$80
(C) \$120
(D) \$160
(E) \$180

6 Tina has half as many sourballs as Louise does. If Louise ate 3 of her sourballs and lost 2 more, she would have one more sourball than Tina does. How many sourballs does Tina have?

(A) 2
(B) 3
(C) 5
(D) 6
(E) 7

7 In a bag of jellybeans, $\frac{1}{3}$ are cherry and $\frac{1}{4}$ are licorice. If the remaining 20 jellybeans are orange, how many jellybeans are in the bag?

(A) 12
(B) 16
(C) 32
(D) 36
(E) 48

HARD

8 If the circumference of a circle is equal to twice its area, then the area of the circle equals

(A) 2
(B) π
(C) 2π
(D) 4π
(E) 16π

9 If $r = \frac{6}{3s+2}$ and $tr = \frac{2}{3s+2}$, then $t =$

(A) $\frac{1}{4}$
(B) $\frac{1}{3}$
(C) 2
(D) 3
(E) 4

10 If x^2 is added to $\frac{5}{4y}$, the sum is $\frac{5+y}{4y}$. If y is a positive integer, which of the following is the value of x?

(A) $\frac{1}{4}$ (B) $\frac{1}{2}$ (C) $\frac{4}{5}$ (D) 1 (E) 5

Answers and Explanations: Problem Set 3

EASY

1 **B** Start with C: $6 = x$. That gives you $6 + 12 = 36$. No good. At this point, don't stare at the other choices waiting for divine inspiration—just pick another one and try it. It's OK if the next answer you try isn't right either. If we plug in 4 for x, we get $4 + 12 = 16$. The equation works, so that's that.

2 **B** Start with C. If the highest number is 6, the other two are 5 and 4. 5 and 4 don't add up to 3—cross out C. Try B: if the highest number is 3, the other two numbers are 1 and 2. (They have to be consecutive.) The sum of $3 + 2 + 1= 6$, and twice the sum of $6 = 12$. If you picked A , you didn't pay attention to what the question asked for. Be sure to reread the question so you know which number they want.

3 **B** Try C first. Does $2^8 = 8^4$? Nope. (Use your calculator.) Try something lower, like B: Does $2^6 = 8^2$? Yessiree.

MEDIUM

4 **C** Try C first. If Jane has \$47 to spend, $47 \div 3 = 15$ with 2 left over. (The shirts cost \$15 apiece.) Now try $47 \div 10 = 4$, with 7 left over. (Socks are \$4 a pair.) It works.

5 **B** Try C first: if Josh spent \$120, Bob spent \$240 and Ralph spent \$40. That means Bob and Ralph together spent \$280, not \$180 as the problem tells us. C is no good. Since our number is way too big, let's try something smaller. If Josh spent \$80, Bob spent \$160 and Ralph spent \$20. So Bob and Ralph together spent \$180. That's more like it.

6 **D** Start with C: if Tina has 5 sourballs, then Louise has 10. If Louise eats 3, then she has 7. If she loses 2 more, she's down to 5. We're supposed to end up with Louise having 1 more than Tina, but they both have 5. Cross out C—and you know you're close to the right answer. Try D: if Tina has 6, Louise has 12. If Louise eats and loses 5, she's got 7, which is one more than Tina has.

7 **E** Try C first. Oops—$\frac{1}{3}$ of 32 is a fraction. Forget C. Try D: $\frac{1}{3}$ of 36 = 12. $\frac{1}{4}$ of 36 = 9. Does 12 + 9 + 20 = 36? No. Try E. $\frac{1}{3}$ of 48 = 16. $\frac{1}{4}$ of 48 = 12. Does 16 + 12 + 20 = 48? Yes!

Making a simple chart will help you keep track of your work:

	D	E
cherry	12	16
licorice	9	12
orange	20	20
TOTAL	41	48

HARD

8 **B** Try C first. If the area is 2π....oops, that makes the radius a fraction. Move on. Try B. If the area is π, then the radius is 1. ($\pi r^2 = \pi$, $r^2 = 1$, $r = 1$.) If $r = 1$, the circumference is $2\pi(1) = 2\pi$. So the circumference is twice the area. Beautiful.

9 **B** Try C first. If $t = 2$, then look at the second equation:

$$2r = \frac{2}{3s+2}$$

$$r = \frac{2}{3s+2} \bullet \frac{1}{2}$$

$$r = \frac{1}{3s+2}$$

Compare that to the first equation. No good. Try B. If $t = \frac{1}{3}$, then

$$\frac{r}{3} = 2(3s + 2)$$

$$r = \frac{2}{3s+2} \bullet 3$$

$$r = \frac{6}{3s+2}$$

Same as the first equation. You're done. Whew.

10 **B** Choice C is particularly nasty here, so let's ignore it. Try Choice D, plugging in 1 for x. You get $1 + \frac{5}{4y} = \frac{5+y}{4y}$ or $1 = \frac{1}{4}$. (The y drops out.) B gives you $\frac{1}{4} + \frac{5}{4y} = \frac{5+y}{4y}$ or $\frac{1}{4} = \frac{1}{4}$. You could also solve by plugging in—choose a positive integer for y, plug it into the equation, and see what happens. You end up with $x = \frac{1}{2}$.

TIP: When you're backsolving, start with C unless C is nasty and hard to work with, as in Question 10. In that case, try the integers, since they'll be easier to do anyway. And don't worry if you have to try a couple of answer choices before you hit the right one—the first one you do is always the slowest, because you're still finding your way. Subsequent tries should be easier. And backsolving is always easier than writing equations.

PROBLEM SET 4: MORE BACKSOLVING

EASY

1 If $\frac{a-4}{28} = \frac{1}{4}$, then $a =$

(A) 11 (B) 10 (C) 7 (D) 6 (E) $\frac{3}{28}$

2 If the area of $\Delta\, ABC$ is 21, and the length of the height minus the length of the base equals 1, then the base of the triangle is equal to

(A) 1 (B) 2 (C) 4 (D) 6 (E) 7

3 If $d^2 = \sqrt{4} + d + 10$, then $d =$

(A) –2 (B) 2 (C) 3 (D) 4 (E) 16

MEDIUM

4 If $\frac{4}{x-1} = \frac{x+1}{2}$, which of the following is a possible value of x?

(A) –1
(B) 0
(C) 1
(D) 2
(E) 3

5 The product of the digits of a two-digit number is 6. If the tens' digit is subtracted from the units' digit, the result is 5. What is the two-digit number?

(A) 61
(B) 32
(C) 27
(D) 23
(E) 16

6 If $16{,}000 = 400(x + 9)$, what is the value of x?

(A) 391
(B) 310
(C) 40
(D) 31
(E) 4

7 What is the radius of a circle with an area of $\frac{\pi}{4}$?

(A) 0.2
(B) 0.4
(C) 0.5
(D) 2
(E) 4

HARD

8 If 20 percent of x is 36 less than x percent of $x - 70$, what is the value of x?

(A) 140
(B) 120
(C) 110
(D) 100
(E) 50

9 If $x^2 = y^3$ and $(x - y)^2 = 2x$, then y could equal

(A) 64
(B) 16
(C) 8
(D) 4
(E) 2

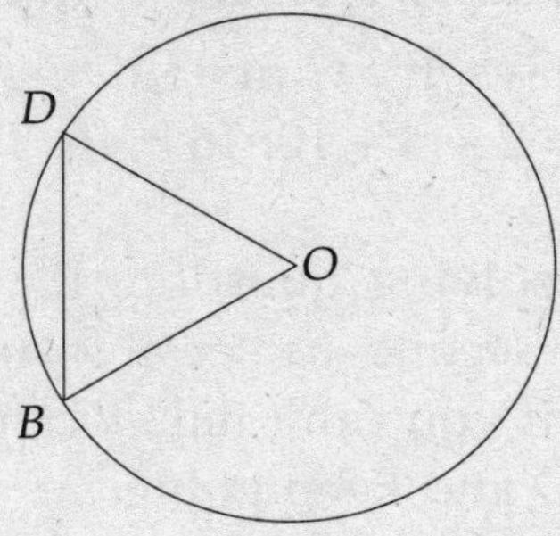

10 In the figure above, $OD = DB$ and arc $DB = 2\pi$. What is the area of the circle?

(A) 64π
(B) 36π
(C) 16π
(D) 12π
(E) 4π

Answers and Explanations: Problem Set 4

EASY

1 **A** Try C first. Does $\frac{3}{28} = \frac{1}{4}$? Nope. Look for a bigger number—B gives us $\frac{6}{28}$, which is closer, but still no cigar. Choice A gives us $\frac{7}{28} = \frac{1}{4}$.

2 **D** Try C first. If the base = 4, then $h - 4 = 1$ and $h = 5$. The formula for area of a triangle is $\frac{1}{2}bh$, so the area would be 10. Too small. Try D: if the base is 6, then $h - 6 = 1$ and $h = 7$. The area is $\frac{42}{2} = 21$. Bingo. If you couldn't do this question without looking up the formula for the area of a triangle, that should tell you something.

3 **D** Yes, it looks nasty, but it's a breeze with the miracle of backsolving. As always, try C first: $3^2 = \sqrt{4} + 3 + 10; 9 = 2 + 13$. Forget it. Try D: $4^2 = 2 + 4 + 10; 16 = 16$. That's it.

TIP: On any backsolving question, if you try C, and it doesn't work, take a second to see if you need a higher or lower number. But if you can't tell *quickly*, don't get hypnotized—just try B or D and keep going.

MEDIUM

4 **E** Try C first. If $x = 1$, does $\frac{4}{0}$ . . . forget it. You can't divide by 0. Not ever. Period. Try D: If $x = 2$, does $\frac{4}{1} = \frac{3}{2}$? No way. Try E. If $x = 3$, $\frac{4}{2} = \frac{4}{2}$. Yes. Remember to avoid trying negatives (like choice A) unless they're all you have left, or you have some reason to think they'll be right.

5 **E** Take the directions of the problem one at a time. The product of the digits = 6, so cross out C. Now for step two. Subtract the tens' digit from the units' digit, and look for 5. E does it. If you picked A, you subtracted the units' digit from the tens' digit, which means you don't know the definitions (see the definitions review at the beginning of the Arithmetic section) or—even more horribly—you didn't reread the question to see what your next direction was. **Always reread the question before continuing on to the next step.** We are human, after all.

6 **D** Try C first. Does $400 \bullet 49 = 16{,}000$? No, and hopefully you just estimate that and don't bother doing it out, with or without your calculator. How about D? $400(40) = 16{,}000$. Yep. If you picked A, you miscounted the zeros. Try checking your answers on your calculator.

7 **C** Fabulous backsolving question. Try C first. Let's convert 0.5 to a fraction, because we like fractions better than decimals and because the question has a fraction in it. If the radius is $\frac{1}{2}$, the area is $\pi\left(\frac{1}{2}\right)^2 = \pi\left(\frac{1}{4}\right) = \frac{\pi}{4}$. Stick a fork in us, we're done.

TIP: Why do we like fractions better than decimals? Mostly because that irritating little decimal point is so easily misplaced. Also because decimals can get very tiny and hard to estimate. You don't want to convert decimals to fractions automatically—only when the question would be easier to do that way. If the question is in decimals and the answers are in decimals, then don't bother converting.

HARD

8 **B** Try D first, because the question is about percents and 100 is easy to do. 20% of 100 is 20. 100% of 100 – 70 is 30. Does 30 – 20 = 46? Nah. Try B: 20% of 120 is 24. 120% of 50 is 60. Does 60 – 24 = 36? Yes.

9 **D** Try C first. If $y = 8$, then $x^2 = 8^3$. $8^3 = 512$. If $x^2 = 512$, x isn't an integer. Forget C. Try D: if $y = 4$, then $x^2 = 4^3$. $x^2 = 64$; $x = 8$. Now try them in the second equation: $(8 - 4)^2 = 2(8)$. $4^2 = 16$. It works. Notice that when C didn't work, we went with a smaller number, because it was easier.

10 **B** First, write in 2π beside arc *DB*. Now try C: if the area is 16π, the radius is 4. Write in 4 beside the two radii, and also *DB*, because *OD* = *DB*. Aha! That makes triangle *DOB* equilateral! Since angle *DOB* is 60°, and $\frac{60}{360} = \frac{1}{6}$, that makes arc *DB* $\frac{1}{6}$ of the circumference. Remember our radius is 4, so the circumference is 8π. Uh oh— 2π is not $\frac{1}{6}$ of 8π. So cross off C. But at least now we know what to do. Try B: if the area is 36π, the radius is 6 and the circumference is 12π. $\frac{1}{6}$ of 12π is 2π. Yeah! Did that seem really painful? It was a lot of work, but then, it was a hard question. The reason backsolving is a good technique for this problem is that if you backsolve it, you get to move through the question like a robot, one step after the other, and you don't have to depend on a flash of insight.

PROBLEM SET 5: ESTIMATING

EASY

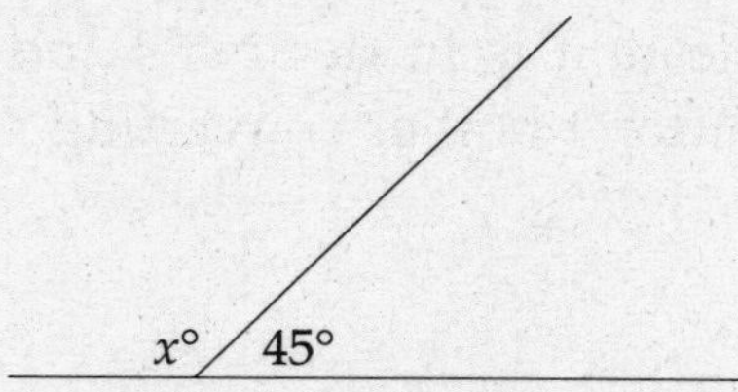

1 What is the value of $2x$?

(A) 360
(B) 270
(C) 135
(D) 90
(E) 67.5

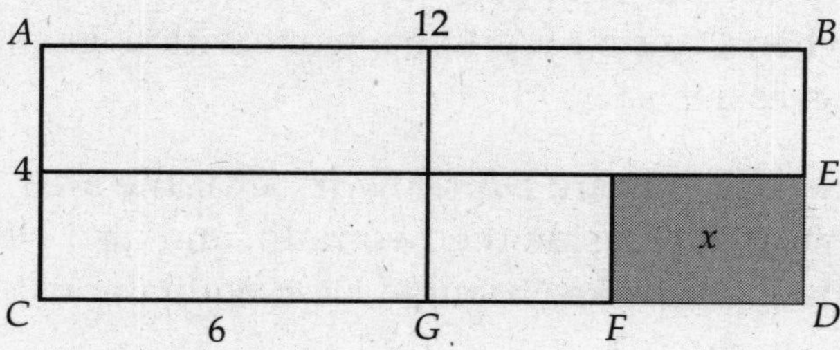

2 If F is equidistant from G and D, and E is equidistant from B and D, what fractional part of rectangle $ABCD$ is area x?

(A) $\frac{1}{16}$ (B) $\frac{1}{8}$ (C) $\frac{1}{4}$ (D) $\frac{1}{3}$ (E) $\frac{1}{2}$

3 If Sarah bought 12 pies for \$30, how many pies could she have bought for \$37.50 at the same rate?

(A) 9
(B) 10
(C) 12
(D) 14
(E) 15

MEDIUM

4 If a runner completes one lap of a track in 64 seconds, approximately how many *minutes* will it take her to run 40 laps at the same speed?

(A) 25
(B) 30
(C) 43
(D) 52
(E) 128

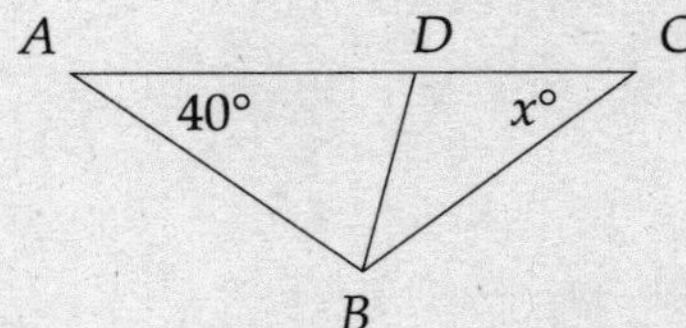

5 In the figure above, $BD = DC$ and $AB = AD$. What is the value of x?

(A) 110
(B) 70
(C) 55
(D) 35
(E) 15

6 Martina wants to buy as many felt-tip pens as possible for $10. If the pens cost between $1.75 and $2.30, what is the greatest number of pens Martina can buy?

(A) 4
(B) 5
(C) 6
(D) 7
(E) 8

7 1.2 is what percent of 600?

(A) 0.002%
(B) 0.2%
(C) 5%
(D) 20%
(E) 500%

HARD

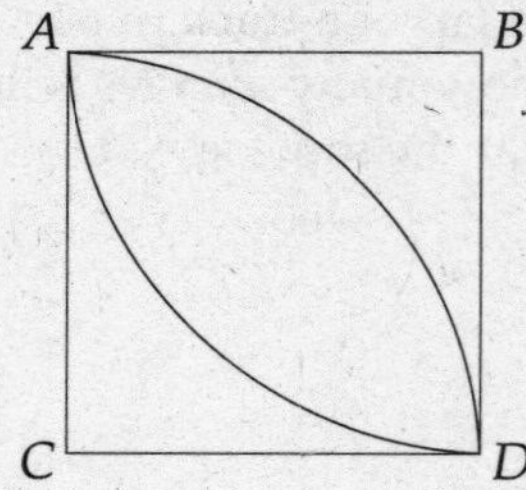

8 In the figure above, $ABCD$ is a square with sides of 4. What is the length of arc AD?

(A) 8π (B) 4π (C) 3π (D) 2π (E) π

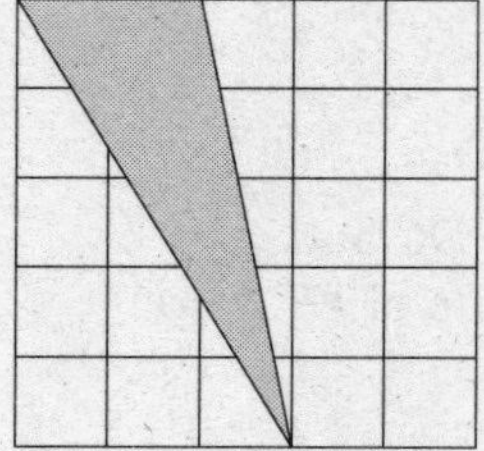

9 Each of the small squares in the figure above has an area of 4. If the shortest side of the triangle is equal in length to 2 sides of a small square, what is the area of the shaded triangle?

(A) 160 (B) 40 (C) 24 (D) 20 (E) 16

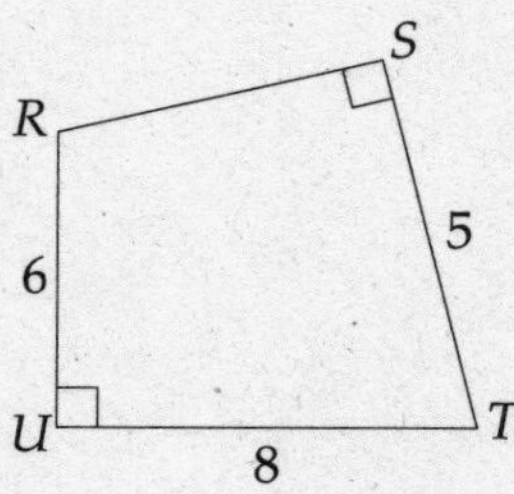

Note: Figure not drawn to scale.

10 In the figure above, what is the length of RS?

(A) 10 (B) $5\sqrt{3}$ (C) 8 (D) $\sqrt{5}$ (E) $2\sqrt{3}$

Answers and Explanations: Problem Set 5

EASY

1 **B** Estimate first: x looks pretty big, doesn't it? Bigger than 90? Yes. So $2x$ will be bigger than 180. Cross out C, D, and E. Now figure out exactly what x is. $x + 45 = 180$, so x is 135. $2x = 270$.

2 **B** Use your eyeballs, and compare against the answer choices. Does x look like $\frac{1}{2}$ of the rectangle? No? Cross out E. What about $\frac{1}{3}$? Cross out D. $\frac{1}{4}$? Cross out C. Could you fit 16 x's in the rectangle? No—cross out A. It's also helpful to draw more boxes in the figure, and then you could count them up:

3 **E** \$37.50 is going to buy more pies than \$30, right? So cross out anything less than or equal to 12. Say goodbye to A, B, and C. Now use your calculator, and divide 30 by 12. You get 2.50—that's the cost per pie. Divide 37.5 by 2.50 and you get 15, which is the number of pies Sarah can buy for \$37.50. You could also set up a proportion if you want:

$$\frac{\text{pies}}{\$}: \frac{12}{30} = \frac{x}{37.50}$$

TIP: Why bother to estimate if you're going to do the whole question anyway? Two reasons: if you eliminate answers using your common sense, you won't make a nutty careless error and wind up picking an answer that's *definitely* wrong. The other reason is that on harder questions, you may not be able to figure out the answer exactly, so you want to be good at estimating so you can guess aggressively and intelligently.

MEDIUM

4 C The runner takes a little over a minute to run one lap, so it will take her a little over 40 minutes to run 40 laps. The proportion would be:

$$\frac{1}{64} = \frac{40}{x}$$

So $x = 2{,}560$. Now convert the seconds to minutes by dividing 2,560 by 60. You get 42.666. Isn't estimating a lot faster?

5 D First just eyeball the angle. It's smaller than 90°. It's close to angle *BAD*, which is marked 40. You're down to C and D. Is it a little smaller than *BAD* or 15° bigger than *BAD*? Go for it! You can always come back and check your answer the long way if you have time.

The long way: *BAD* is isosceles, since AB = AD. The two base angles of *BAD* =140, so each is 70°. If $\angle BDA = 70°$, then $\angle BDA$ is 110°. Triangle *BDC* is isosceles too, with $\angle DCB$ and $\angle CBD = x°$. $2x = 70$, so $x = 35°$. Now admit it—isn't estimating easier?

6 B If Martina wants to buy as many pens as possible, she wants to buy the cheapest ones she can. Try backsolving—with C, if she buys 6 for $1.75, that equals 10.50. A bit over $10, so pick the next lowest answer choice.

7 B 600 is pretty big, and 1.2 is pretty tiny. So you should be looking for a pretty small percentage. Cross out D and E, and maybe even C. Then use your calculator. The easiest way to figure out your next step would be to set up a proportion:

$$\frac{1.2}{600} = \frac{x}{100}$$

Or transform the sentence:

$$1.2 = \frac{x}{100} \bullet 600$$

HARD

8 D First mark the sides of the square with 4. Now estimate the length of *AD*, based on the side of the square. Think of the side of the square as a piece of spaghetti that you are going to drape over *AD*. So *AD* is longer than 4. Maybe around 6? Now go to the answers, and substitute 3 for π. (We know, $\pi = 3.14$, but you don't have to be so exact. We're just estimating.) Choice A is around 24. Way too big. B is around 12, C is around 9, and D is around 6. E is too small. We'd pick D and move on.

9 D When you estimate, remember that each shaded square has an area of 4. It's tricky to do this exactly, because mostly only slivers of squares are shaded. So fake it. Almost 2 full squares at the top, another square on the next row (that's 3 so far) and then slivers on the next 3 rows that make up about 2 full squares. So we've got 5 squares each with area 4; the area of the triangle is around 20. Hey, let's pick D and take a nap.

OK, OK, so you want to cross out A and B and then get the answer exactly? We can do that. If each square has an area of 4, then the side of a little square is 2. Write that on the figure, in a couple of places. Now use the top of the triangle as the base. It equals 4. The other thing we need is the height, or altitude, of the triangle—and in this case, the height is equal to a side of the big square, or 10. Using the formula for the area of a triangle (we know you didn't have to look that up—we just know it), plug 4 in for the base and 10 in for the height and you get 20 for the area.

10 B Hey, wake up!! You can't estimate anything if the figure isn't drawn to scale! But you may want to re-draw the figure to make it look more like it's supposed to look. Now for the solution: Draw a line from *R* to *T*, slicing the figure into 2 triangles. Now all you have to do is use the Pythagorean Theorem to calculate the lengths. Triangle *RUT* is a 6-8-10 triangle, a Pythagorean triple. Now for *RST*: $a^2 + 5^2 = 10^2$. So $a^2 = 75$ and $a = 5\sqrt{3}$.

PROBLEM SET 6: QUANT COMP

EASY

	Column A	Column B
1	3(2 – 1)	2(3 – 1)

$x > 0$

	Column A	Column B
2	$\frac{x^2}{x}$	$\frac{x^3}{x^2}$

	Column A	Column B
3	The number of hours in 5 days	The number of minutes in 3 hours

$2y - 4 = 10$
x is positive

	Column A	Column B
4	$8y - x$	$y^2 - x$

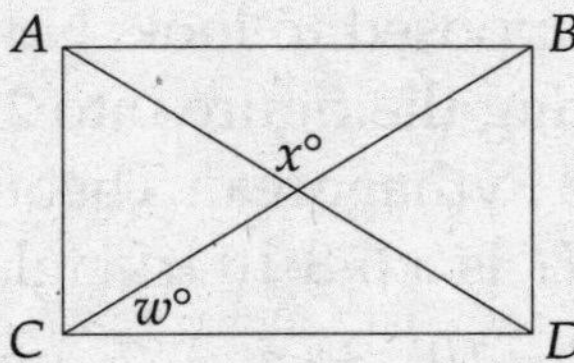

ABCD is a rectangle.

	Column A	Column B
5	$2w$	$180 - x$

MEDIUM

	Column A	Column B
6	The volume of a box that is 8 inches high, 10 inches wide, and x inches long	The volume of a box that is 20 inches high, high, y inches wide, and 4 inches long

$3x + 4y = 18$
$x + y = 4$

	Column A	Column B
7	$4x + 5y$	20

$y < 0$
y is an integer.

	Column A	Column B
8	$\frac{2y}{3}$	$\frac{3y}{2}$

	Column A	Column B
9	The number of distinct letters used in Sentence P	The number of distinct words used in Paragraph Q

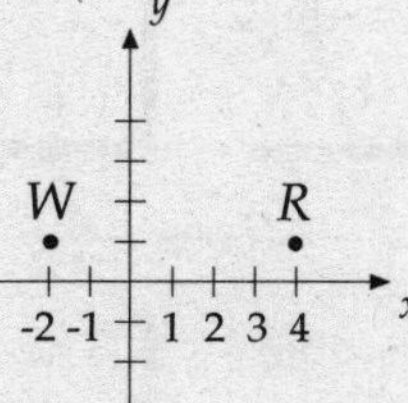

	Column A	Column B
10	The shortest distance from Point W to a point on the y-axis	The shortest distance from Point R to a point on the x-axis

HARD

	Column A	Column B

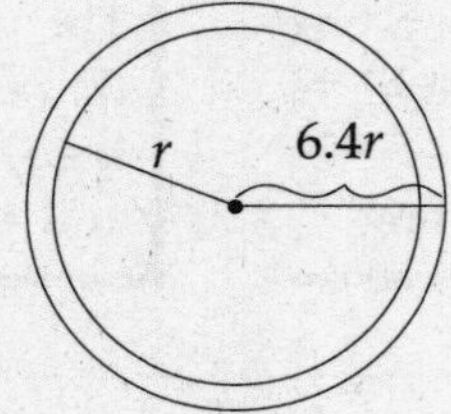

Note: Figure not drawn to scale.

	Column A	Column B
11	The circumference of the smaller circle	The radius of the larger circle

$1 < x < 5$

	Column A	Column B
12	$4x$	x^2

A box of cookies contains only oatmeal cookies and chocolate wafers.

	Column A	Column B
13	The number of chocolate wafers in the box if the ratio of oatmeal cookies to chocolate cookies wafers is 3:2	The difference in the number of chocolate wafers and oatmeal in the box

	Column A	Column B
14	The greatest possible solution to the equation $x^2 - x - 6 = 0$	The greatest possible solution to the equation $x^2 + x - 6 = 0$

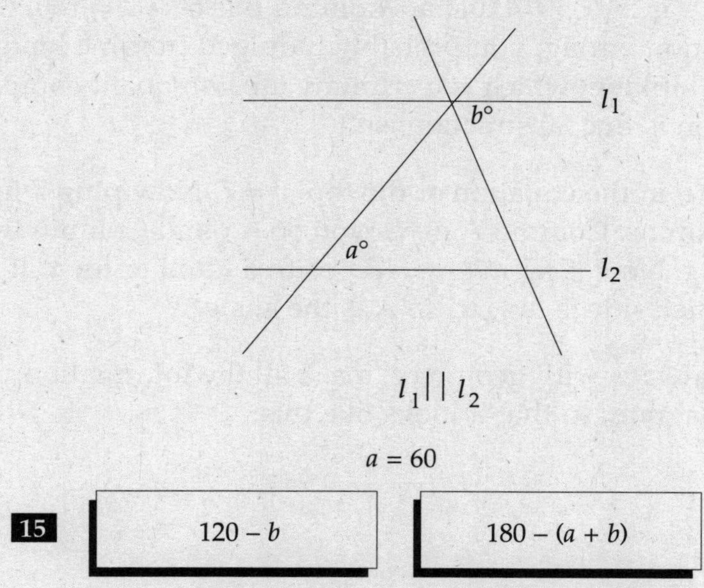

$l_1 \| l_2$

$a = 60$

15	$120 - b$	$180 - (a + b)$

Answers and Explanations: Problem Set 6

On all quant comp questions, write ABCD next the problem and cross each choice out until you're left with one answer. That way you'll be more accurate. If you picked E on anything, give yourself a spanking.

EASY

1 **B** Don't forget the order of operations—PEMDAS. Parentheses first. So Column A is 3 • 1 = 3 and Column B is 2 • 2 = 4.

2 **C** First reduce $\frac{x^2}{x} = x$, and $\frac{x^3}{x^2} = x$. Review the exponents rules if you need to. If you forgot the rule, you could plug in a number for x, something bigger than 0. How about 1? (On quant comp, it's OK to use 1. Sometimes using 1 is the trick to getting the right answer.) If $x = 1$, then Column A is $\frac{1}{1}$ and Column B is $\frac{1}{1}$. Try $x = 2$: Column A is $\frac{4}{2}$ and Column B is $\frac{16}{8}$. Still equal.

3 **B** Column A is 24 • 5 = 120, and Column B is 60 • 3 = 180. If you got this question wrong, you probably indulged in some kind of abstract thinking, which is a crummy idea on quant comp. Estimation is fine, abstraction isn't.

4 **A** Solve for y in the equation at the top. $y = 7$. Now plug 7 in for y in both Columns. Column A gives you $56 - x$ and Column B gives you $49 - x$. Now plug in any old positive number for x. It won't affect which side is bigger, so A is the answer.

5 **C** First, as always with geometry, mark all the information you know on the diagram, so that it looks like this:

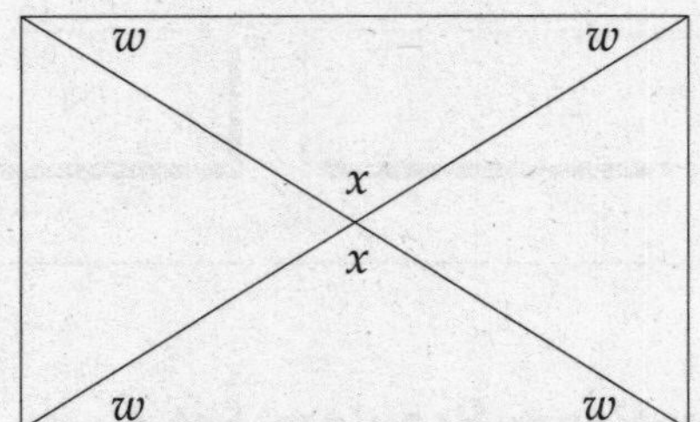

Now you see that the bottom triangle has angles measuring w, w, and x. Plugging in will make this easier: $w = 20$ and $x = 140$. Column A is 40 and Column B is 40. Beautiful.

MEDIUM

6 **D** You're missing the length in Column A, and the width in Column B. Without those measurements, we can't figure anything out. If you don't believe it, try plugging in numbers.

7 **A** Hey! Don't ever solve for x and y individually when you have simultaneous equations, unless you've tried adding or subtracting equations first. If you add 'em together you get $4x + 5y = 22$. That's better, isn't it?

8 **A** Plug in a negative number for y. (The directions say you have to.) Plug in $x = -2$. Column A is $-\frac{4}{2}$ and Column B is $-\frac{6}{2}$. Be careful with negatives—if you got this far and picked B, you forgot that the closer to 0, the bigger the negative number. If you miss a lot of these questions, just make a number line and plot your numbers on it, and you won't go wrong.

9 **D** If we don't know what the sentence is, we don't know how many letters are in it. Same goes for the paragraph. Maybe it's short. Maybe it's long. Who knows?

10 **A** First draw your lines in on the diagram. It should look like this:

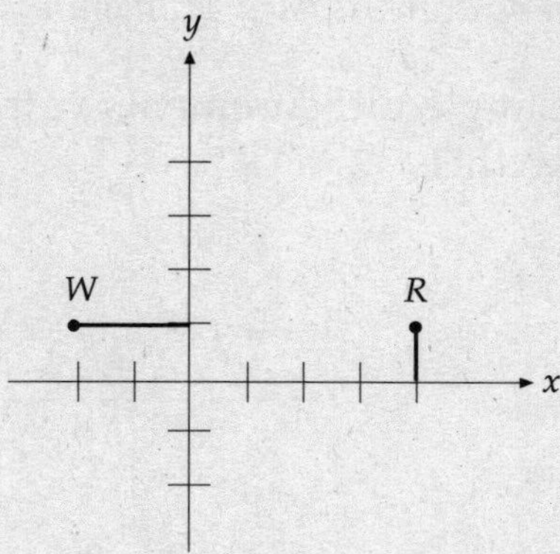

Which one is longer?

HARD

11 **B** Plug in $r = 2$. That makes the circumference of the smaller circle 4π, and the radius of the larger circle 12.8. 4π is just a bit over 12, so B is bigger. If you're uncertain, try plugging in a different number. **Bad guess:** A, because A looks bigger, but the figure isn't drawn to scale.

12 **D** Plug in $x = 2$. Column A is 8, Column B is 4. Cross out B and C. Now make x bigger. If $x = 4$, Column A and Column B are equal. **Bad guess: B.** Just because x has an exponent doesn't mean it's necessarily bigger. On hard questions the answer won't be that easy to get.

13 **D** Column A tells us the ratio, but we don't have any idea how many cookies there are. Maybe 3 chocolate and 2 oatmeal. Or maybe 15 chocolate and 10 oatmeal. Column B is even less helpful—we don't even have a ratio, so we don't have any idea how many of each we have.

14 **A** Don't try to eyeball this one. Factor each equation. Column A factors to $(x - 3)(x + 2) = 0$, so its greatest possible solution is 3. Column B factors to $(x + 3)(x - 2) = 0$, so its greatest possible solution is 2. **Bad guess:** B. Just because B has a positive term and A has a negative term, that doesn't automatically make B bigger. *Don't forget you're doing a hard problem.*

15 **C** Mark in all the angle measurements you can. Your diagram should look like this:

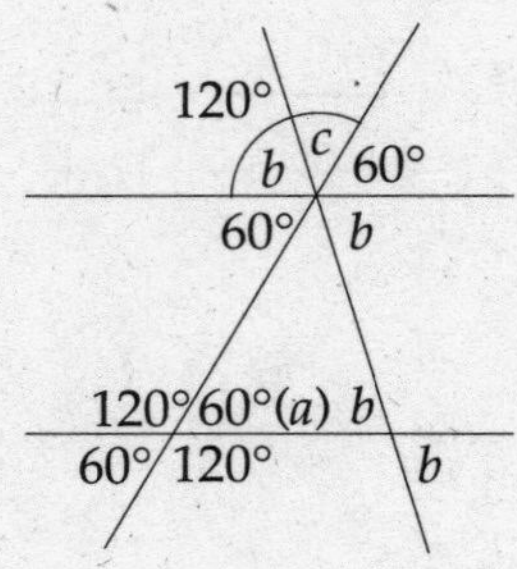

What do you know now? $a + b + c = 180$, so $180 - (a + b) = c$. And $b + c = 120$, so $120 - b = c$. **Bad guess:** D. It looks ugly, doesn't it? Then don't guess D.

PROBLEM SET 7: MORE QUANT COMP

EASY

	Column A	Column B
1	The number of days in three months	The number of hours in four days

$x = -1$

2	$x - x$	0

3	$-40 + 29 + 53 - 71$	$29 - 71 + 40 + 53$

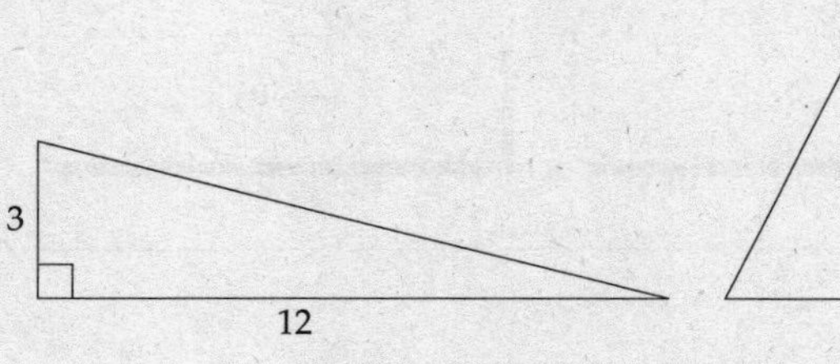

Triangle A Triangle B

4	The area of A	The area of B

Each session of Seminar B starts at 8 AM and ends at 3 PM on the same day.

5	The number of hours in one session of Seminar B	5

MEDIUM

	Column A	Column B
	$2x = y$ x and y are integers.	
6	x	y
	Dora ate 6 more crackers than Bob did. Bob ate 4 fewer crackers than Jeff did. Jeff ate c crackers.	
7	The number of crackers Dora ate	$c + 2$
	$x^2 + 2x - 15 = 0$	
8	The greatest possible value of x	5
	$6 - n < -1$	
9	$-n$	$-(-n)$

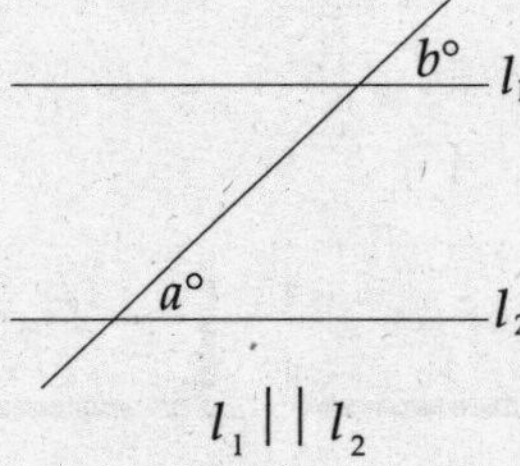

Note: Figure not drawn to scale.

	Column A	Column B
10	$180 - a$	$180 - b$

HARD

Column A	Column B

$$\frac{2}{x} = \frac{y}{3}$$

11 | xy | $2y$ |

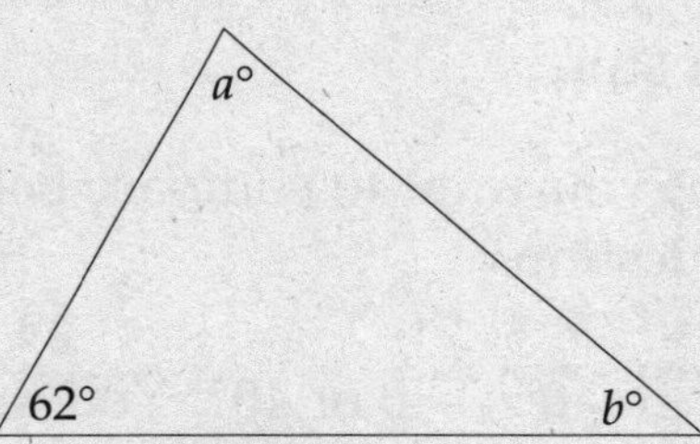

Note: Figure not drawn to scale.

12 | b | $2a$ |

Cube C has an edge of 1 inch.

13 | The number of square inches in the surface area of Cube C | The number of cubic inches in the volume of Cube C |

In Triangle ABC, $AB = 6$ and $BC = 4$

14 | The length of AC | 2 |

The "flip" of a number is defined as the reciprocal of the number multiplied by the square root of the number.

15 | The "flip" of $\frac{1}{4}$ | The "flip" of 1 |

Answers and Explanations: Problem Set 7

EASY

1 **B** What is the greatest possible number of days in three months? $31 \times 3 = 93$. Compare this to the number of hours in four days. $24 \times 4 = 96$.

2 **C** Always write it out, don't do it in your head. Column A is $-1 - (-1) = -1 + 1 = 0$. It's *very* easy to make a careless mistake on this kind of stuff.

3 **B** Remember that you want to compare before you calculate in easy quant comp questions

$$-40 + \cancel{29} + \cancel{53} + \cancel{71} \qquad \cancel{29} - \cancel{71} + 40 + \cancel{53}$$

Now which is greater, –40 or 40? You can also use your calculator in this question, but it takes a little more time and requires you to be very careful.

4 **C** Just figure out the areas of each—Triangle A is $\left(\frac{1}{2}\right)(12)(3) = 18$, and Triangle B is $\left(\frac{1}{2}\right)(6)(6) = 18$.

5 **A** Whoa! Just count on your fingers: the number of hours from 8 to 3 is 7. If you have a watch with a face (not digital) just count them up by going around the dial.

MEDIUM

6 **D** Plug in 1 for x, which makes $y = 2$. Eliminate answers A and C. Now plug in 2 for x, which makes $y = 4$. Hmm. Column B is still bigger. How about $x = 0$, which makes $y = 0$. Now Column A and Column B are equal, so the answer is D.

7 **C** Plug in numbers. If $c = 5$, then Bob ate 4 and Dora ate 7. Column A is 7 and Column B is 7. You don't have to worry about negatives because you can't eat a negative number of crackers. If you want to try another set of numbers, go ahead.

8 **B** Factor the equation: $(x + 5)(x - 3) = 0$. That means the greatest possible value of x is 3. (To figure that out, just set each of the parentheses equal to 0. So in this equation, $x = -5$ or 3.)

9 **B** Be careful when you manipulate the inequality. You get $-n < -7$. If you divide through by -1, you get $n > 7$. (The sign changes direction when you multiply through by a negative number.) Now let's plug in 8 for n. Column A is -8 and Column B is 8. **Bad guess:** A. Just because B looks like it's full of negatives, that doesn't mean it's necessarily smaller.

10 **C** As always, write everything you know on the diagram. Since you have parallel lines, even though the diagram isn't drawn to scale, $a = b$. ("Figure not drawn to scale" only means the picture may not look the way it's supposed to. It *doesn't* mean they're lying when they give you info—such as degree measurements, or that 2 lines are parallel.)

HARD

11 **D** Try plugging in. Let $x = 3$ and $y = 2$. Column A is now 6 and Column B is now 4. Cross out answers B and C. Try another set of numbers: $x = 1$ and $y = 6$. Column A is now 6 and Column B is now 12. So you can't tell.

12 **D** We know that $a + b + 62 = 180$, so $a + b = 118$. But how are those degrees divided between a and b? We don't know. If $a = 10$, then $b = 108$ and Column A is bigger. If $b = 10$, then $a = 108$ and Column B is bigger. **Bad guess:** B. You can't estimate when the figure isn't drawn to scale.

13 **A** To get the surface area of a cube, first get the area of one face, then multiply by 6 (because all 6 faces are equal). One face of Cube *C* has an area of 1, so the surface area is 6. The volume of a cube is s^3, so the volume of Cube *C* is 1.

14 **A** The rule here is that any 2 sides of a triangle must add up to more than the third side. (Otherwise you'd have a straight line—and it doesn't matter whether your triangle is a right triangle or not.) So the smallest leg possible in *ABC* has to bigger than 2, because $2 + 4 = 6$, and according to the rule, $AC + 4 > 6$.

15 **A** Hello—it's a function question. So just follow the directions. In Column A, the reciprocal of $\frac{1}{4}$ is $\frac{4}{1}$, and the square root of $\frac{1}{4}$ is $\frac{1}{2}$. So the "flip" of $\frac{1}{4}$ is $4 \times \frac{1}{2} = 2$. In Column B, the reciprocal of 1 is 1, and the square root of 1 is 1. So the "flip" of 1 is 1. **Bad guess:** B. Oh please. You think because 1 is bigger than $\frac{1}{4}$ that's all there is to it? *Don't fall asleep on hard questions.*

PROBLEM SET 8: FACTORS, MULTIPLES, & PRIMES

EASY

1 If t is even, which of the following expressions must be odd?

(A) $t - 2$
(B) t^2
(C) $2(t + 1)$
(D) $t(t + 1)$
(E) $t + 3$

2 If m is a multiple of 5 and n is a factor of 3, which of the following could equal 13?

(A) mn

(B) $m + n$

(C) $\frac{m}{n}$

(D) $\frac{n}{m}$

(E) $m - n$

Set A: {0, 1, 2, 3, 4, 5}
Set B: {1, 2, 7, 9, 10}

3 How many member of Set A are factors of any member of Set B?

(A) 2
(B) 3
(C) 4
(D) 5
(E) 6

MEDIUM

4 Which of the following equations is equal to $6y + 6x = 66$?

(A) $33 = x + y$
(B) $33 = 2y + 2x$
(C) $11 - x = y$
(D) $11 - 2x = y$
(E) $4y - 4x = 44$

5 If the greatest prime factor of 32 is a, and the least prime factor of 77 is b, then ab is divisible by which of the following numbers?

(A) 3
(B) 4
(C) 8
(D) 11
(E) 14

6 $[x]$ is defined as the greatest prime factor of x minus the least prime factor of x. What is the value of $\frac{[20]}{[10]}$?

(A) 10
(B) 5
(C) 2
(D) 1
(E) $\frac{1}{2}$

7 If p is the number of prime numbers between 65 and 75, then $p =$

(A) 0
(B) 1
(C) 2
(D) 3
(E) 4

HARD

8 If r, s, and t are positive integers and $rs = t$, then which of the following must be true?

I. $r < t$
II. $r \leq t$
III. $s \geq t$

(A) I only
(B) II only
(C) III only
(D) I and III only
(E) II and III only

9 If $12y = x^3$, and x and y are positive integers, what is the least possible value for y?

(A) 3
(B) 9
(C) 18
(D) 27
(E) 64

10 The alarm of Clock A rings every 4 minutes, the alarm of Clock B rings every 6 minutes, and the alarm of Clock C rings every 7 minutes. If the alarms of all three clocks' alarms ring at 12:00 PM, the next time at which all the alarms will ring at exactly the same time is

(A) 12:28 PM
(B) 12:56 PM
(C) 1:24 PM
(D) 1:36 PM
(E) 2:48 PM

Answers and Explanations: Problem Set 8

EASY

1 **E** Plug in an even number for t. How about $t = 2$? Now go to the answers, looking for something to be odd. A: 2, B: 4, C: 6, D: 10, and E: 5. And remember that anything multiplied by an even number will be even, so you can eliminate B, C, and D.

2 **B** Plug in again. Let's make $m = 10$ and $n = 3$. Run those through the answer choices, looking for 13. Choice B is $m + n$, or 10 + 3 or 13. If you picked other numbers and didn't get an answer, don't get frustrated or think you're doing something wrong. Just pick another set of numbers, and make sure your numbers fit the directions of the problem, (i.e., the number you pick for m has to be a multiple of 5, and the number you pick for n has to be either 1 or 3).

3 **C** Let's be methodical about this. Take each number in Set A, one at a time, and see if it divides evenly into anything in Set B. 0 isn't a factor of anything but 0. 1 is a factor of everything in Set B. Put a check by 1. 2 is a factor of 2, so put a check by 2. 3 is a factor of 9, so put a check by 3. 4 isn't a factor of anything in Set B. 5 is a factor of 10, so put a check by 5. How many checks do we have? Four of 'em.

MEDIUM

4 **C** Try reducing the equation in the question first, by dividing the whole thing by 6. That leaves you with $x + y = 11$. That's the same as C, if you just move the x to the other side.

5 **E** There's only one prime factor of 32, and it's 2. The prime factors of 77 are 7 and 11, so the smallest is 7. That makes $ab = 14$, which is certainly divisible by 14.

6 **D** This is a function question, as we're sure you noticed. First deal with [20]. The greatest prime factor of 20 is 5, and the least prime factor is 2. 5 – 2 = 3, so [20] = 3. Now for [10]. The greatest prime factor of 10 is 5, and the least prime factor is 2. So [10] = 5 – 2 = 3. $\frac{3}{3} = 1$. If you picked C, you tried to weasel out of doing the function. You won't get away with it.

7 **D** First write out the numbers: 66, 67, 68, 69, 70, 71, 72, 73, 74. Be methodical—cross out anything that's even: 66, 68, 70, 72, 74. Now cross out anything left that's divisible by 3: 69. Any number divisible by 4, 6, or 8 is even, and we've already crossed those out. Any number divisible by 9 is also divisible by 3, and we're crossed that out. All we have left is 7—we're left with 67, 71, and 73. None of them is divisible by 7, so they're all prime.

HARD

8 **B** Plug in. Let's make a little chart:

r	*s*	*t*
1	2	2
1	1	1
2	3	6

That should do it. Now check our numbers against I, II, and III. I isn't true if r, s, and t all equal 1. III isn't true if $r = 2$, $s = 3$, and $t = 6$. That leaves us with II.

Why bother with a chart? On *must be* questions, picking one set of numbers probably isn't going to be enough. This question is pretty tricky for a medium, because I is true except when all the variables equal 1. But then I and II only isn't a choice, so you have to disprove one of them.

9 **C** Solve this problem with a combination of factoring and backsolving. The question looks like this: $2 \times 2 \times 3y = x^3$. Now factor the answer choices. Choice A is $3 \bullet 1$. If you plug that in for y, does it give you a cube? Nope. B is $3 \bullet 3$. No good either. C is $2 \times 3 \times 3$—now we have $(2 \times 3)(2 \times 3)(2 \times 3) = x^3$. It works.

Another way to do this problem is more straightforward backsolving: plug in the answers for y, and use your calculator to see if that product is a cube root. If your calculator doesn't have the x^y function, then make a list of cubes to see if the product is on it.

10 C Ouch—this one is ugly. You can't simply multiply $4 \times 6 \times 7$ and add 168 minutes to 12:00. You'll get E, and while it's true that all 3 alarms would ring at 2:48, that's not the *earliest* time they would ring at the same time. (And that solution is too easy for a hard question.) Instead, factor the ringing rates, so you get (2×2), (2×3) and (7). The lowest common multiple will be $2 \times 2 \times 3 \times 7$, 4 goes in evenly, and so do 6 and 7. Now multiply it, and you get 84, which is 1 hour and 24 minutes. Add that to 12:00, and you're done.

TIP: When you're backsolving remember that if the question asks for the *least possible value*, start with the smallest answer choice. For *greatest possible value*, start with the biggest answer choice. That way you won't get caught picking an answer that works, but isn't the *least* or *greatest* answer that works.

PROBLEM SET 9: FRACTIONS, DECIMALS, PERCENTS

EASY

1 A big-screen TV is on sale at 15% off the regular price. If the regular price of the TV is $420, what is the sale price?

(A) $63
(B) $126
(C) $357
(D) $405
(E) $435

2 Which of the following is the decimal form of $70 + \frac{7}{10} + \frac{3}{1000}$?

(A) 70.0703
(B) 70.7003
(C) 70.703
(D) 70.73
(E) 77.003

3 Six more than two thirds of twelve is

(A) 10
(B) 12
(C) 14
(D) 18
(E) 22

MEDIUM

4 Walking at a constant rate, Stuart takes 24 minutes to walk to the nearest bus stop, and $\frac{1}{3}$ of that time to walk to the movie theater. It takes him half the time to walk to school that it does for him to walk to the movie theater. How many minutes does it take Stuart to walk to school?

(A) 36
(B) 24
(C) 16
(D) 8
(E) 4

5 What is the value of x if $\frac{\frac{1}{2}}{x} = 4$?

(A) 8 (B) 2 (C) $\frac{1}{2}$ (D) $\frac{1}{4}$ (E) $\frac{1}{8}$

6 If x% of y is 10, then y% of x is

(A) 1
(B) 5
(C) 10
(D) 50
(E) 90

7 A certain drink is made by adding 4 parts water to 1 part drink mix. If the amount of water is doubled, and the amount of drink mix is quadrupled, what percent of the new mixture is drink mix?

(A) 30%

(B) $33\frac{1}{3}$%

(C) 50%

(D) $66\frac{2}{3}$%

(E) 80%

HARD

8 Set A consists of distinct fractions, each of which has a numerator of 1 and a denominator d such that $1 < d < 8$, where d is an integer. If Set B consists of the reciprocals of the fractions with odd denominators in Set A, then the product of Set A and Set B =

(A) $\frac{1}{96}$
(B) $\frac{1}{48}$
(C) $\frac{1}{24}$
(D) 1
(E) 8

9 For all values x, if x is even, x^* is defined as $0.5x$; if x is odd, x^* is defined as $\frac{x}{3}$. What is the value of $\frac{(6a)^*}{9^*}$?

(A) $\left(\frac{2a}{3}\right)^*$
(B) $\left(\frac{a}{3}\right)^*$
(C) a^*
(D) $(2a)^*$
(E) $(3a)^*$

10 If a, b, and c are distinct positive integers and 10% of abc is 5, then $a + b$ could equal

(A) 1
(B) 3
(C) 5
(D) 6
(E) 25

Answers and Explanations: Problem Set 9

EASY

1 C The numbers are too awkward to backsolve, so do it the old-fashioned way: 15% of \$420 is $0.15 \times 420 = 63$. $420 - 63 = 357$. Use your calculator.

2 C Take the pieces one at a time and eliminate. The first piece is 70: eliminate E. The second piece is $\frac{7}{10}$, or 0.7. Eliminate A. The last piece is $\frac{3}{1000}$, or 0.003. Eliminate B and C. If you want to do the conversions on your calculator, that's cool. But adding the fractions together and then converting to a decimal would be a massive waste of your precious time, calculator or no calculator.

3 C Translate the problem into math language: $6 + \frac{2}{3} \bullet 12 =$
Then, don't forget PEMDAS: multiply before you add.
$6 + \frac{2}{3}(12) = 6 + 8 = 14$

MEDIUM

4 E Start working from the 24 minutes it takes poor Stuart to walk to the bus stop. (Won't anybody give the guy a ride?) If it takes $\frac{1}{3}$ of 24 to walk to the movies, that's 8 minutes. If it takes him half of that time to walk to school, $\frac{1}{2}$ of 8 is 4. This question requires close reading more than anything else.

5 E Backsolving wouldn't be a bad idea here—you can eliminate A and B pretty quickly that way. C gives you $\frac{\frac{1}{2}}{\frac{1}{2}} = 1$. D is $\frac{\frac{1}{2}}{\frac{1}{4}} = 2$.
E is $\frac{\frac{1}{2}}{\frac{1}{8}} = 4$.

6 **C** Let's plug in 20 for y, which makes $x = 50$. Plug those numbers into the second part: 20% of 50 = 10. (Spooky, isn't it?)

7 **B** First make a little chart: If you double the water and quadruple the mix, you get

water		mix
4	:	1
8	:	4

Reread the question. It asks for the percentage of the new mixture that's drink mix. We've got $\frac{4\text{ (mix)}}{12\text{ (total)}}$, which equals $\frac{1}{3}$, or $33\frac{1}{3}\%$. If you made it almost to the end but picked C, don't forget that you have to express the mix as a percentage of the total, not a percentage of the water.

HARD

8 **B** Whew. You have to read this carefully. Set A has different fractions, each with a numerator of 1. (You might as well write them down like that, and fill in the denominators when you get there.) The denominators are between 1 and 8. That gives you Set A: $\frac{1}{2}, \frac{1}{3}, \frac{1}{4}, \frac{1}{5}, \frac{1}{6}, \frac{1}{7}$. Set B has the reciprocals of the members of Set A with odd denominators, so Set B: $\frac{3}{1}, \frac{5}{1}, \frac{7}{1}$. Now we're going to multiply the sets together—see how the fractions that have reciprocals cancel each other out? You're left with $\frac{1}{2} \times \frac{1}{4} \times \frac{1}{6}$, which is $\frac{1}{48}$.

9 **D** It's a function, so just follow the directions. Looking at the numerator, $6a$ has to be even because it has an even number as a factor. (Or plug in any low number for a.) Since $6a$ is even, follow the first direction; $6a \times 0.5 = 3a$. Now for the denominator: 9 is odd, so follow the second direction. $\frac{9}{3} = 3$. So $\frac{(6a)^*}{9^*} = \frac{3a}{3} = a$. Did you pick C? Well, sorry, we aren't getting off that easy. The answers are functions too, so we have to translate them, looking for our answer, a. Skip A and B—they look too complicated. For D, $(2a)$ is even, so $(2a)^* = 2a \times 0.5 = a$. Finito.

10 **B** First translate the middle part of the problem into an equation. 10% of abc is 5 translates to $\frac{10}{100} \bullet abc = 5$. Now solve for abc, and you get $abc = 50$. Reread the question. Each variable is different, each is positive, and multiplied together they produce 50. Now backsolve, and remember that the answers represent $a + b$. Choice A is silly, because it would make a and b fractions, and they can't be fractions. In B, $a + b$ would have to be $1 + 2$. If $a = 1$ and $b = 2$ and $abc = 50$, what is c? $c = 25$, so it works.

TIP: A couple of reminders: if you are making mistakes on the easy and medium problems, don't spend a lot of time—if any—working on the hard problems. You need to hone your skills first; you may want to go back to the review section and do some work before continuing. And don't forget, you probably want to leave some questions blank on the real thing.

Speaking of leaving questions blank, Question 9 would be a fine choice to avoid entirely. Long functions in the hard problems can be really nasty.

PROBLEM SET 10: RATIOS, PROPORTIONS, & PROBABILITIES

EASY

1 If $\frac{8}{x} = \frac{2}{3}$, then $x =$

(A) 12
(B) 10
(C) 6
(D) 4
(E) 3

2 A factory produces 6,000 plates per day. If one out of 15 plates is broken, how many unbroken plates does the factory produce each day?

(A) 5800
(B) 5600
(C) 1500
(D) 800
(E) 400

3 It takes 4 friends 24 minutes to wash all the windows in Maria's house. The friends all work at the same rate. How long would it take 8 friends, working at the same rate, to wash all the windows in Maria's house?

(A) 96
(B) 32
(C) 20
(D) 12
(E) 8

MEDIUM

4 In a certain classroom there is an equal number of boys and girls. If 2 girls and 1 boy left to go home, the ratio of boys to girls in the room would be 4:3. How many girls are in the classroom now?

(A) 2 (B) 3 (C) 4 (D) 5 (E) 6

5 Bobby picks 5 blueberries every 2 seconds. How long will it take Bobby to pick 400 blueberries?

(A) 2 minutes, 20 seconds
(B) 2 minutes, 40 seconds
(C) 3 minutes, 10 seconds
(D) 4 minutes, 30 seconds
(E) 8 minutes

6 A drawer holds only blue socks and white socks. If the ratio of blue socks to white socks is 4:3, which of the following could be the total number of socks in the drawer?

(A) 4 (B) 7 (C) 8 (D) 12 (E) 24

7 The probability of choosing a caramel from a certain bag of candy is $\frac{1}{5}$, and the probability of choosing a butterscotch is $\frac{5}{8}$. If the bag contains 40 pieces of candy, and the only types of candy in the bag are caramel, butterscotch, and fudge, how many pieces of fudge are in the bag?

(A) 5
(B) 7
(C) 8
(D) 16
(E) 25

HARD

8 The ratio of $\frac{1}{6}:\frac{1}{5}$ is equal to the ratio of 35 to

(A) 24
(B) 30
(C) 36
(D) 42
(E) 45

9 An artist makes a certain shade of green paint by mixing blue and yellow in a ratio of 3:4. She makes orange by mixing red and yellow in a ratio of 2:3. If on one day she mixes both green and orange, and uses equal amounts of blue and red paint, what fractional part of the paint that she uses is yellow?

(A) $\frac{7}{12}$
(B) $\frac{17}{29}$
(C) $\frac{7}{5}$
(D) $\frac{17}{12}$
(E) $\frac{9}{6}$

10 The areas of two circles are in a ratio of 4:9. If both radii are integers, and $r^1 - r^2 = 2$, which of the following is the radius of the larger circle?

(A) 4
(B) 5
(C) 6
(D) 8
(E) 9

Answers and Explanations: Problem Set 10

EASY

1 **A** Cross-multiply, and you get $24 = 2x$ so $x = 12$.

2 **B** First estimate. You're looking for the number of unbroken plates—if only one broke out of 15, there should be a lot of unbroken plates, right? Cross out C, D, and E. Now set up a proportion:

$$\frac{\text{broken}}{\text{total}} = \frac{1}{15} = \frac{x}{6000}$$

And cross-multiply. You get $6000 = 15x$, so using your calculator, $x = 400$. That's the broken plates, so subtract 400 from 6000 and you've got the answer. If you picked E, you could have gotten the problem right if you had either estimated first or reread the question right before you answered it.

3 **D** There are twice as many people, so the work will go twice as fast. You can't set up a normal proportion because it's an inverse proportion—the more people you have, the less time the work takes. So if you multiply the number of people by 2, you divide the work time by 2. Don't forget to use your common sense.

MEDIUM

4 **D** Let's backsolve, shall we? Try C first. If there are 4 girls, there are 4 boys, since the numbers start out equal. Now subtract 1 boy and 2 girls and you get 3:2. No good. Try D: 5 boys, 5 girls, 1 boy and 2 girls leave—that gives us 4:3. Isn't backsolving fab?

5 **B** Set up a proportion:

$$\frac{\text{berries}}{\text{seconds}} = \frac{5}{2} = \frac{400}{x}$$

Cross-multiply to solve for x, and $x = 160$ seconds. You can estimate from here—160 seconds is well over minutes. Pick B and keep moving. To convert it exactly, divide 160 by 60. You get with a remainder of 40. Halt! Danger! Don't convert the remainder to a fraction or a decimal when you converting seconds/minutes/hours. In this case, the remainder is the number of seconds left over, and you need it: the answer is 2 minutes, 40 seconds.

6 **B** The total must be the sum of the numbers in a ratio, or a multiple of that sum. In this case, 4 + 3 = 7, so the number of socks could be 7 or any multiple of 7. (You can have fractions in a ratio, it's true, but not when you're dealing with socks or people or anything that you can't chop into pieces. And probably not on a medium question either.)

7 **B** Okey-dokey. Here's what to do: take $\frac{1}{5}$ of 40, which is 8 caramels. Take $\frac{5}{8}$ of 40, which is 25 butterscotches. The caramels and the butterscotches are 8 + 25 = 33. Subtract that from 40 and you've got the fudge.

HARD

8 D First multiply the ratio by something big, to get rid of the fraction. Any multiple of 6 and 5 will do. So $30\left(\frac{1}{6}\right):30\left(\frac{1}{5}\right) = 5:6$. Now we've got 5:6 = 35:x. Since 35 is 5 × 7, x is 6 × 7, or 42. The new ratio is 35:42, which is the same as 5:6.

9 B Write down your ratios and label them neatly. You have

$$\frac{b:y}{3:4} \quad \frac{r:y}{2:3}$$

If the artist uses equal amount of blue and red, we have to multiply each ratio:

$$\frac{b:y}{(2)(3:4)} \quad \frac{r:y}{(2:3)(3)}$$

The result is

$$\frac{b:y}{6:8} \quad \frac{r:y}{6:9}$$

The yellow is 8 parts + 9 parts = 17 parts, and the total is 6 + 8 + 6 + 9 = 29 parts. It's important, on complicated ratio problems, to organize the information legibly and label everything as you go along, or else you'll find yourself looking at a bunch of meaningless numbers.

10 C Backsolve! Start with C: if the larger radius is 6, the smaller radius is 2 less than that, or 4. Area of smaller circle = 16π, and area of larger circle is 36π. 16π:36π is a ratio of 4:9. (Just divide the whole ratio by 4π.) If you picked E you must've had a momentary blackout—that answer is way too appealing to be right on a hard question. If you're going to guess, guess something less obvious.

PROBLEM SET 11: AVERAGES

EASY

1 Three consecutive integers add up to 258. What is the smallest integer?

(A) 58
(B) 85
(C) 86
(D) 89
(E) 94

2 If the average (arithmetic mean) of x, $2x$, and 15 is 9, then x =

(A) 2
(B) 4
(C) 7
(D) 8
(E) 9

a	1, 2, 3, 4, 5
b	21, 22, 23, 24, 25

3 What is the mean of the sum of the members of a + the sum of the numbers of b?

(A) 13
(B) 18
(C) 21
(D) 25
(E) 26

MEDIUM

4 The average (arithmetic mean) of 4 numbers is 36. If the two middle numbers add up to 62, what is the average of the smallest and largest numbers?

(A) 20
(B) 36
(C) 41
(D) 72
(E) 82

Questions 5–6 refer to the following definition.

For all numbers x and y, $(x, y)!$ is defined as the average of x and y.

5 $(2, 10)! - (2, 2)! =$

(A) $(2, 5)!$
(B) $(2, 10)!$
(C) $(3, 6)!$
(D) $(5, 4)!$
(E) $(7, 1)!$

6 Which of the following represents $(a, b, c)!$?

(A) $3abc$

(B) abc

(C) $\frac{abc}{2}$

(D) $\frac{a+b+c}{3}$

(E) $\frac{3}{a+b+c}$

7 Dixie spent an average of x dollars on each of 5 shirts, and an average of y dollars on each of 3 hats. In terms of x and y, how many dollars did she spend on shirts and hats?

(A) $5x + 3y$
(B) $3x + 5y$
(C) $15(x + y)$
(D) $8xy$
(E) $15xy$

HARD

8 Fred played a total of 7 rounds of golf, and the 7 scores he received were consecutive integers. If Fred's lowest score was g, then in terms of g, what was his average score for all 7 rounds?

(A) $\frac{g+7}{7}$

(B) $\frac{7(g+1)}{g}$

(C) $7g + 7$

(D) $g + 3$

(E) $g + 7$

9 A certain car can travel on a highway for 300 miles on 12 gallons of gas, and in a city for 270 miles on 15 gallons of gas. If the car uses an equal number of gallons for highway driving and city driving, what is the car's average number of miles per gallon?

(A) 15
(B) 18.5
(C) 20
(D) 21.5
(E) 25

10 On 5 math tests, Gloria had an average score of 86. If all test scores are integers, what is the lowest average score Gloria can receive on the remaining 3 tests if she wants to finish the semester with an average score of 90 or higher?

(A) 90
(B) 92
(C) 94
(D) 96
(E) 97

Answers and Explanations: Problem Set 11

EASY

1 **B** If the three integers add up to 258, then their average is 258 ÷ 3, or 86. Since the integers are consecutive, they must be 85, 86, and 87. Check it on your calculator. If you picked C, what did you do wrong? Forget what the question asked for? Divide 258 by 3 and then quit? Even easy problems may have more than one step. This would also be a good question to backsolve.

2 **B** Backsolve. Try C first: if $x = 7$, is the average of 14, 7, and 15 equal to 9? Doesn't that average look too big? Try B: Is the average of 4, 8, and 15 equal to 9? You bet.

3 **A** Time for an adding trick: first add up a and get 15. For b, add the tens' digits and the units' digits separately: $5 \times 20 = 100$, and you already know that $1 + 2 + 3 + 4 + 5 = 15$, because you just added it. So $b = 100 + 15 = 115$. Now add a and b to get 130, and divide by the number of numbers in both sets, which is 10. And of course, you should use your calculator. When you have a lot of numbers to punch in, be sure to look at each one as it appears in the window so that if you punch in 10 instead of 100 you catch it. Estimating will help too.

MEDIUM

4 **C** If the average of the numbers is 36, then their sum is $4 \times 36 = 144$. If the 2 middle numbers equal 62, subtract them from the sum, and $144 - 62 = 82$. 82 is the sum of the remaining numbers, so to get their average, divide by 2.

5 **E** All we have to do is take the average of the 2 numbers in the parentheses. (2, 10)! is the average of 2 and 10, which is 6. (2, 2)! is the average of 2 and 2, which is 2. $6 - 2 = 4$. Circle 4. Now we have to translate the answer choices, since they are functions as well. E is the average of 7 and 1, which is 4.

6 **D** To take the average of a set of numbers, add them up and divide by the number in the set. So all we have to do is add a, b, and c and divide by 3.

7 **A** Plug in. Let $x = 2$. If Dixie spent an average of \$2 a shirt, then she spent a total of \$10 on shirts. Let $y = 4$, and she spent an average of \$4 a hat, for a total of \$12. Our total is 10 + 12 = \$22. Circle that. Now on to the answer choices, and $x = 2$ and $y = 4$. Choice A is 5(2) + 3(4) = 22. Bingo.

HARD

8 **D** Plug in. Let's say Fred scored a 2 on his first round, so $g = 2$. (We know that's an impossible golf score, but who cares? We want a low number to work with.) If his scores were consecutive numbers, his scores would be 2, 3, 4, 5, 6, 7, 8. The average of those is 5. Circle 5. Plug in 2 for g in the answers and you get D: 2 + 3 = 5.

9 **D** Figure the highway first. 300 miles on 12 gallons is 300 ÷ 12 = 25 miles per gallon. In the city, 270 ÷ 15 = 18 miles per gallon. Since the car uses an equal number of gallons on the highway and in the city, we can average the miles per gallon for each, and the average of 25 and 18 is 21.5.

10 **E** If Gloria averaged 86 on 5 tests, her total number of points so far is 86 × 5 = 430. If she wants to get an average of 90 on 8 tests, she needs a grand total of 90 × 8 = 720 points. So dear Gloria has 3 tests to make up the difference of 720 and 430, which is 290. If you divide 290 by 3, you get 96.66—so, she needs to get an average of 97 on the remaining 3 tests.

TIP: A shortcut for averages: if the list of numbers is consecutive, consecutive odd, or consecutive even, then the average will be the middle number. (If the list has an even number of elements, you have to average the 2 middle numbers.) The average will also be the middle number (or average of the 2 middle numbers) of any list that goes up in consistent increments. For example, the average of 6, 15, 24, 33, and 42 is 24, since the numbers go up in increments of 9.

TIP: One more thing: notice how the right answers to hard algebra questions don't look very appealing—in Question 8, didn't you think the right answer would have a 7 in it? This is why plugging in works so well—you aren't tempted to think in abstract terms and get tricked into picking the wrong answer.

PROBLEM SET 12: EXPONENTS AND ROOTS

EASY

1 If $t^3 = -8$, then $t^2 =$

(A) –4
(B) –2
(C) 2
(D) 4
(E) 8

2 If $r + \sqrt{r} = s^2 - 6$ and $r = 25$, then $s =$

(A) 5 (B) 6 (C) 7 (D) 9 (E) 19

3 If $3^x = 27$, then $4^x =$

(A) 8 (B) 12 (C) 16 (D) 64 (E) 128

MEDIUM

4 For all integers x and y, let

$\bigstar\,(x + y) = \dfrac{x^2}{y^2}$. What is the value of

$\bigstar\,(2 + y) \times \bigstar\,(y + 1)$?

(A) 16
(B) 9
(C) 5
(D) 4
(E) 3

5 If p and k are integers, and $\sqrt{p} = 3^3\sqrt{k}$, then p must be

(A) odd
(B) even
(C) positive
(D) negative
(E) greater than 1

6 $\frac{\sqrt{a} \bullet \sqrt{b}}{3\sqrt{a} - 2\sqrt{a}} =$

(A) $\frac{\sqrt{b}}{\sqrt{a}}$

(B) $\sqrt{b}$

(C) $\frac{2\sqrt{a}}{b}$

(D) $\sqrt{ab}$

(E) $\sqrt{a^2 b}$

7 If $a^5 = 16a^3$, then $a =$

(A) 2

(B) 4

(C) 8

(D) 16

(E) 32

HARD

8 If $0 > a^3bc^6$, then which of the following must be true?

I. ab is positive
II. ab is negative
III. abc is negative

(A) I only
(B) II only
(C) III only
(D) I and III only
(E) II and III only

9 For positive integers p, t, x, and y, if $p^x = t^y$ and $x - y = 3$, which of the following CANNOT equal $\sqrt{t}$?

(A) 1 (B) 2 (C) 4 (D) 9 (E) 25

10 If $\frac{4y}{k}$ is the cube of an integer greater than 1, and $k^2 = y$, what is the least possible value of y?

(A) 1 (B) 2 (C) 4 (D) 6 (E) 27

Answers and Explanations: Problem Set 12

EASY

1 **D** $t = -2$, and $(-2)^2 = 4$.

2 **B** Plug 25 in for r and solve:

$25 + \sqrt{25} = s^2 - 6$

$30 = s^2 - 6$

$36 = s^2$

$6 = s$

3 **D** $x = 3$, and $4^3 = 64$. How did we know $x = 3$? Just plug in—
$3^2 = 9$, $3^3 = 27$. There you go.

MEDIUM

4 **D** All you have to do is square the first thing and put it over the square of the second thing. So $(2 + y) = \frac{2^2}{y^2}$. And $(y + 1) = \frac{y^2}{1}$. Now multiply them. The y^2 cancels so you get 2^2.

5 **C** Let's plug in. If $p = 1$, then $k = 1$. Cross out B, D, and E. Try $p = 4$. Then the cube root of k is equal to 2, so $k = 8$. That works too. Cross out A. (On the SAT, if you cross out 4 answers just pick what's left and keep going. Don't stop to wonder why.)

6 **B** Remember that you can multiply or divide what's under a square root sign, and add or subtract when what's under the square root sign is the same. Begin by simplifying the denominator.

$$\frac{\sqrt{a} \bullet \sqrt{b}}{3\sqrt{a} - 2\sqrt{a}} = \frac{\sqrt{a} \bullet \sqrt{b}}{\sqrt{a}} = \sqrt{b}.$$

7 **B** Your first step in a manipulating equation problemis to isolate the variable. In this equation, divide both sides by a^3. $\frac{a^5}{a^3} = \frac{16a^3}{a^3}$

Remember that when you divide exponents, you subtract. So you're left with $a^2 = 16$. To solve for a, take the square roots of both sides. $\sqrt{a^2} = \sqrt{16}$; $a = 4$.

HARD

8 **B** First let's analyze our inequality. All the stuff to the right is less than zero, so a^3bc^6 is negative. Anything with an even exponent has to be positive, so we know c^6 is positive. That means in a^3b, one variable is positive and the other negative. Let's go to the Roman numerals. I, we know is *not* true, because either a or b is negative, and the other positive. Cross out A and D. Hey! II is just what we're looking for! Cross out C. Now let's look at III. We know ab is negative—but we don't know a thing about c. We know that c^6 is positive, but c could be positive or negative. So cross out E.

This question is very difficult—you could plug numbers in, but in a question like this, they only get in the way. Your only concern is the sign of the variables.

9 **B** Let's make this as easy on ourselves as we can. Plug in $x = 4$ and $y = 1$. Now let's plug in for p and t, starting as low as we can. If $p = 1$, then $t = 1$, and $\sqrt{1} = 1$. Cross off A. Remember, the question asks for a number that CANNOT equal $\sqrt{t}$. So if you plug in and get an integer for $\sqrt{t}$, cross that answer off. If $p = 2$, then $2^4 = t^1$ and $t = 16$. $\sqrt{16} = 4$. Cross off C. Hey—we skipped over B, and we were plugging in for p in order. So pick B. (If you don't believe it, try $p = 3$ and $p = 5$ and see what happens. Just don't waste time doing it on the SAT.)

10 **C** Backsolve. Since the question asks for the least possible value, start with A: if $y = 1$, $k = 1$. $\frac{4(1)}{1} = 4$, which isn't a cube. Cross off A. B: if $y = 2$, then $k = \sqrt{2}$. Forget it, you don't have an integer. C: if $y = 4$, then $k = 2$. $\frac{4(4)}{2} = 8$, which is the cube of 2. That was pretty painless, wasn't it?

PROBLEM SET 13: EQUATIONS: SIMPLE, QUADRATIC, SIMULTANEOUS

EASY

1 If $x^2 = \sqrt{y} + 2$, and $y = 4$, then $x =$

(A) 2
(B) 3
(C) 4
(D) 8
(E) 36

2 If $60 = (7 + 8)(x - 2)$, then $x =$

(A) 15
(B) 10
(C) 9
(D) 7
(E) 6

3 If $4x - 2y = 10$ and $7x + 2y = 23$, what is the value of x?

(A) $\frac{1}{3}$
(B) 1
(C) 3
(D) 13
(E) 14

MEDIUM

4 If $\frac{x+2}{3} = \frac{(x+2)^2}{15}$, what is one possible value of x?

(A) −1
(B) 0
(C) 1
(D) 2
(E) 3

5 For their science homework, Brenda and Dylan calculated the volume of air that filled a basketball. If the formula for the volume of a sphere is $V = \frac{4}{3}\pi r^3$, and the diameter of the basketball was 6, what was the volume of the air inside the basketball?

(A) 4π
(B) 14π
(C) 32π
(D) 36π
(E) 72π

6 $2x - \frac{4}{3} =$

(A) $\frac{12x-3}{4}$
(B) $\frac{6x-4}{3}$
(C) $\frac{5x-3}{4}$
(D) $\frac{x}{3}$
(E) $\frac{2x}{3}$

7 On a certain test, Radeesh earned 2 points for every correct answer and lost 1 point for every incorrect answer. If he answered all 30 questions on the test and received a score of 51, how many questions did Radeesh answer incorrectly?

(A) 3
(B) 7
(C) 15
(D) 21
(E) 24

HARD

8 If b is an integer and $x^2 + bx - 30 = 0$, which of the following CANNOT be the value of b?

(A) 1
(B) 6
(C) 7
(D) 13
(E) 29

9 Twelve baseball cards were sold at an average price of $3 a card. If some of the cards cost $5 each and the rest cost $2 each, how many $5 cards were sold?

(A) 2
(B) 4
(C) 6
(D) 8
(E) 10

10 If $\frac{a}{b} + a = 6$, what is the value of $\sqrt{\frac{a + ab - 2b}{b}}$?

(A) 2

(B) $\sqrt{6}$

(C) 3

(D) 4

(E) $\sqrt{10}$

Answers and Explanations: Problem Set 13

EASY

1 A The question gives you $y = 4$, so plug it in.

$x^2 = \sqrt{4} + 2$
$x^2 = 2 + 2$
$x^2 = 4$
$x = 2$

2 E Backsolve. Try C first: (15)(9 – 2) = (15)(7) = 105. It should equal 60, so we need a much smaller number. Try E: (15)(6 – 2) = (15)(4) = 60. It works. Or you could solve the equation algebraically:

$60 = 15(x - 2)$
$60 = 15x - 30$
$90 - 15x$
$6 - x$

3 C Stack 'em and add:

$4x - 2y = 10$
$7x + 2y = 23$
$11x = 33$
$x = 3$

MEDIUM

4 E Backsolve, C first, of course. That gives you $\frac{1+2}{3} = \frac{(1+2)^2}{15}$ which is $\frac{3}{3} = \frac{9}{15}$. Well, that didn't work. E gives you $\frac{3+2}{3} = \frac{(3+2)^2}{15}$. $\frac{5}{3} = \frac{25}{15}$ and $\frac{5}{3} = \frac{5}{3}$. That works. If you solve this algebraically you get $x = 3$ or -2, and who needs to waste time getting an answer that isn't even one of the choices? You don't.

5 **D** Don't worry—you weren't supposed to know this formula. That's why they gave it to you, so don't get freaked out. Just use the information in the question to solve for V. If the diameter of the basketball was 6, the radius was 3:

$$V = \frac{4}{3}\pi r^3, \text{ so } V = \frac{4}{3}\pi(3^3) = \frac{4}{3}\pi(27) = 36\pi.$$

You may see some totally unfamiliar formula on the test—physics, for instance—but you don't have to understand the formula, or know anything about it. All you have to do is substitute in any value they give you and solve for the variable they ask for.

6 **A** You could always plug in: if $x = 5$, then $2(5) - \frac{4}{3} = 10 - \frac{4}{3} = \frac{26}{3}$. Circle $\frac{26}{3}$. B is $\frac{6(5)-4}{3} = \frac{26}{3}$. To do the algebra, convert $2x$ to $\frac{6x}{3}$. Now you have $\frac{6x}{3} - \frac{4}{3} = \frac{6x-4}{3}$.

7 **B** Backsolve! If Radeesh got 2 points for every right answer and the test had 30 questions, the top score was 60. If he got a 51 he did pretty well, so start with A. (Remember, the answer choices represent the number of questions he answered incorrectly.) If he missed 3, then he got 27 right. $27 \times 2 = 54$. Subtract 3 for 3 wrong answers, and you get 51. Fabulous.

(Don't worry if you didn't see which answer to start with. If you started with C, it gave you way too many wrong answers, didn't it? So cross off C, D, and E and you've only got 2 left to try.)

HARD

8 **B** Backsolve, of course. Start with C. That makes the equation $x^2 + 7x - 30 = 0$. Can you factor that? Sure— it's $(x + 10)(x - 3)$. Cross out C. Now try B, because 6 is an easy number to work with. (It's better than 13 and 29, isn't it?) Try factoring $x^2 + 6x - 30$. You can't. Since you're looking for the answer that CANNOT be k, B is correct.

9 **B** Backsolve. Try C first: if 6 cards worth \$5 were sold, that's a total of \$30. If the total sold was 12, that leaves 6 cards worth \$2, for a total of \$12. To get the average cost per card, divide the total cost by the number of cards: \$42 ÷ 12 isn't \$3. Try something smaller: B gives you 4 cards at \$5, for a total of \$20, and 8 cards at \$2 for a total of \$16. That adds up to \$36. Now for the average: \$36 ÷ 12 = \$3. The "math" way to solve this would be to write simultaneous equations: $x + y = 12$ and $5x + 2y = 3(x + y)$. What's wrong with doing it that way? Well, first you have to write not one, but two equations correctly. Then you have to solve them. All of that takes too much time, and allows too many opportunities for mistakes.

10 **A** This is a very difficult question. Since there are 2 variables, and the equation is equal to a specific number, we can't plug in. (As you may have found out.) Since the answer choices are equal to that ugly square root with 2 variables, we can't backsolve either. Since there aren't 2 equations, we can't do simultaneous equations. So we're stuck with algebra.

Look at the square root we're solving for. Ignore the square root sign for a minute, and do the division underneath.

$$\frac{a+ab-2b}{b} = \frac{a}{b} + \frac{ab}{b} - \frac{2b}{b} = \frac{a}{b} + a - 2.$$

Now look back at the equation: if $\frac{a}{b} + a = 6$, you can substitute 6 for that part, which leaves you with $\sqrt{6-2} = \sqrt{4} = 2$.

If you didn't get this question right, don't sweat it unless you're aiming for a 700 or over. Even then, this would be a good problem to leave blank.

PROBLEM SET 14: LINES, ANGLES, COORDINATES

EASY

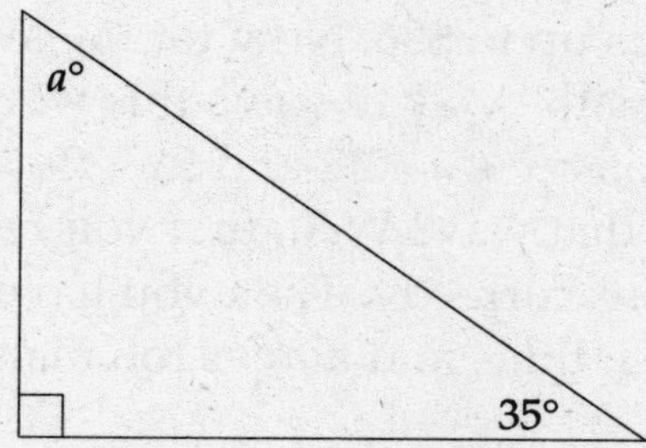

1 In the figure above, $3a - a =$

(A) 40
(B) 55
(C) 90
(D) 110
(E) 165

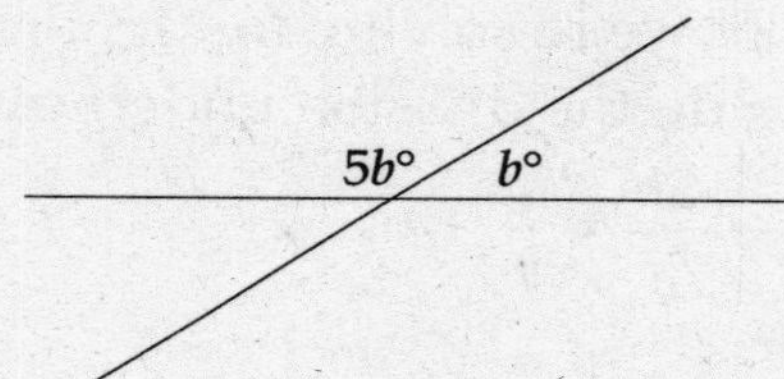

2 In the figure above, $b =$

(A) 20
(B) 30
(C) 40
(D) 45
(E) 180

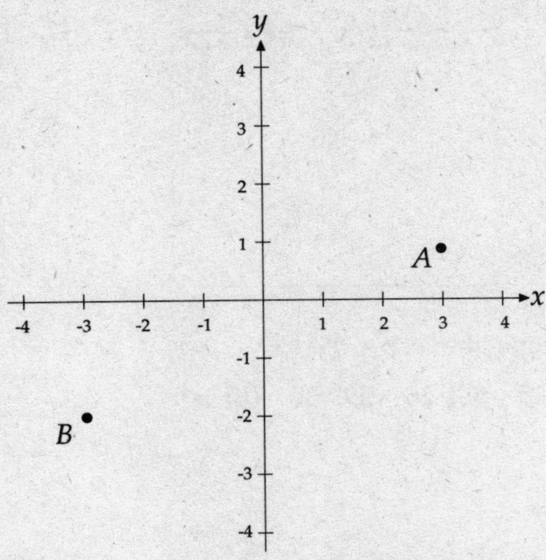

3 The x-coordinate of Point A minus the y-coordinate of Point B equals

(A) –2 (B) –1 (C) 0 (D) 3 (E) 5

MEDIUM

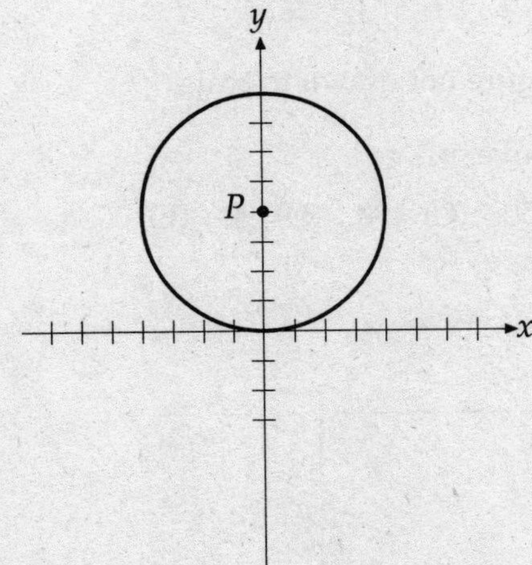

4 Point P is the center of Circle Q, which has a radius of 4. Which of the following points lies on Circle Q?

(A) (4, 0)
(B) (0, 4)
(C) (–4, 4)
(D) (3, 1)
(E) (4, 3)

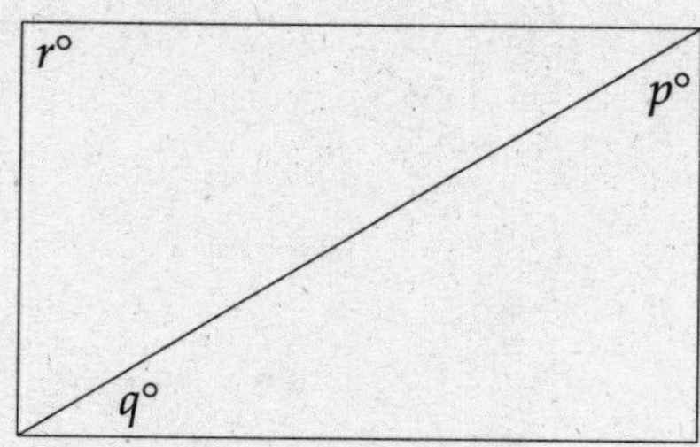

5 In the rectangle above, $p + q - r =$

(A) 0 (B) 15 (C) 26 (D) 35 (E) 50

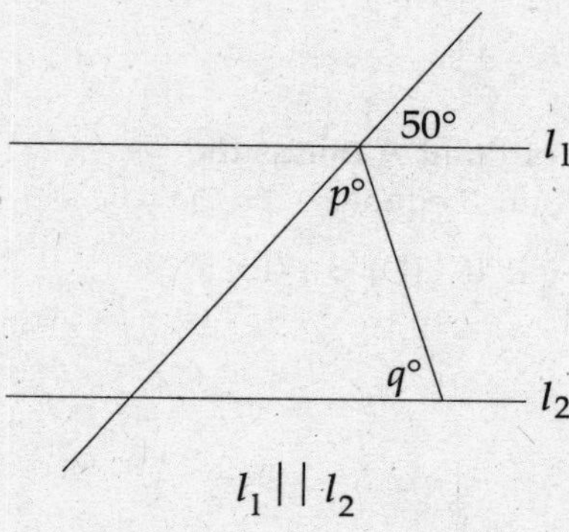

Note: Figure not drawn to scale.

6 In the figure above, $p + q =$

(A) 180 (B) 150 (C) 130 (D) 90 (E) 70

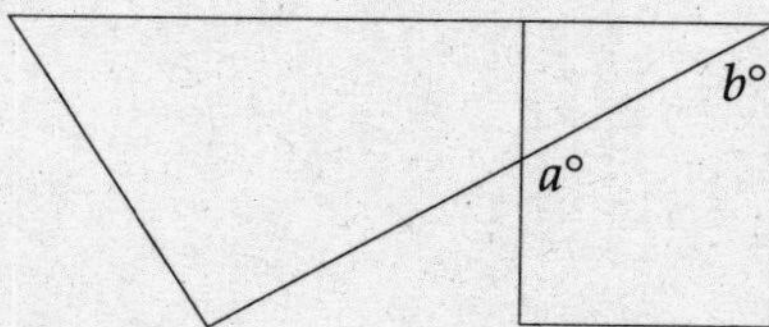

7 The figure above is formed by a triangle overlapping a rectangle. What does $a + b$ equal?

(A) 80 (B) 90 (C) 150 (D) 180 (E) 270

HARD

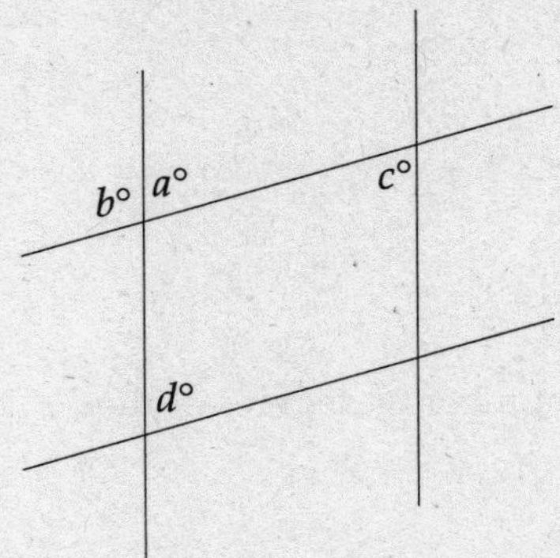

Note: Figure not drawn to scale.

8 Which of the following statements must be true?

I. $a + b < 180$
II. $a + d = 180$
III. $a + d > 180$

(A) None
(B) I only
(C) II only
(D) I and II only
(E) II and II only

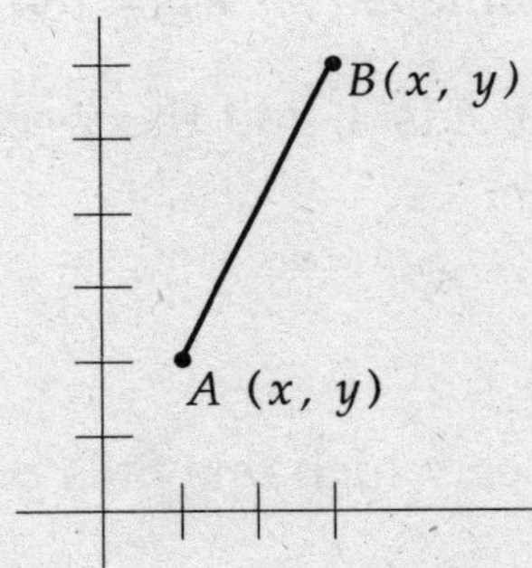

9 If the coordinates of Point A are (1, 2), what is the slope of line AB?

(A) –3

(B) –2

(C) $\frac{1}{2}$

(D) 2

(E) 3

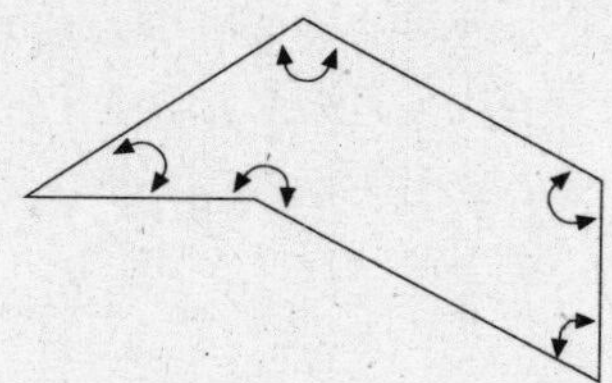

10 What is the total number of degrees of the marked angles?

(A) 180
(B) 270
(C) 360
(D) 540
(E) 720

Answers and Explanations: Problem Set 14

EASY

1 **D** A triangle has 180°. So $90 + 35 + a = 180$, and $a = 55$. Plug that into the equation and get $3(55) - 55 = 110$.

2 **B** $5b$ and b lie on a straight line, so $5b + b = 180$ and $b = 30$.

3 **E** The x-coordinate of Point A is 3, and the y-coordinate of Point B is -2. So $3 - (-2) = 5$.

MEDIUM

4 **C** Just plot the points and see which one falls on the circle.

5 **A** As always, mark whatever info you can on your diagram: $r = 90$, and the bottom angle is also 90, because this is a rectangle. $p + q = 90$, because they are the 2 remaining angles in a right triangle. That means $p + q - r = 0$.

6 **C** Again, mark info on the diagram. The unmarked angle of the triangle is 50°, and so $50 + p + q = 180$, and $p + q = 130°$. (You can't figure out what p and q are individually, but the question doesn't ask you to.)

7 **D** Estimate first. a is around 130 and b is bigger than 45, so $a + b$ should be a little bigger than 175. Pick D and keep cruising. To figure the angles exactly, ignore the triangle and look at the quadrilateral in the bottom half of the rectangle. The angles are $a + b + 90 + 90$. Since a quadrilateral has 360°, $a + b = 180$.

HARD

8 **A** It's important to realize what you *don't* know: are any of these lines parallel? *We don't know.* So we can't draw any conclusions at all. None. Zero, zip, nada.

9 **D** Mark the points on your diagram, so that (x, y) is $(1, 2)$. That makes $(3x, 3y) = (3, 6)$. The slope is $\frac{6-2}{3-1} = \frac{4}{2} = 2$.

10 **D** Estimate first, and see what you can cross out. Since you have 2 angles that are bigger than 90 and 1 angle that's bigger than 180, you should be able at least to cross out A, B, and C. To figure out the exact number of degrees, divide the figure into 3 triangles:

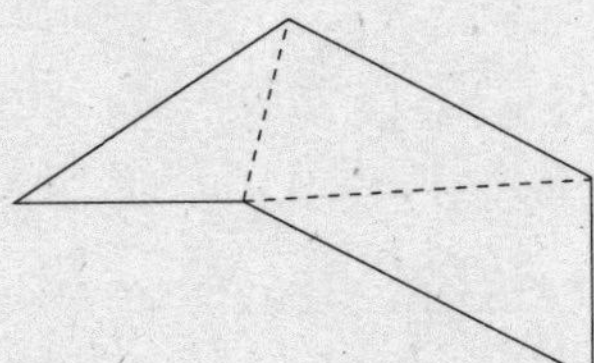

The total degrees will be $180 \times 3 = 540$.

PROBLEM SET 15: TRIANGLES

EASY

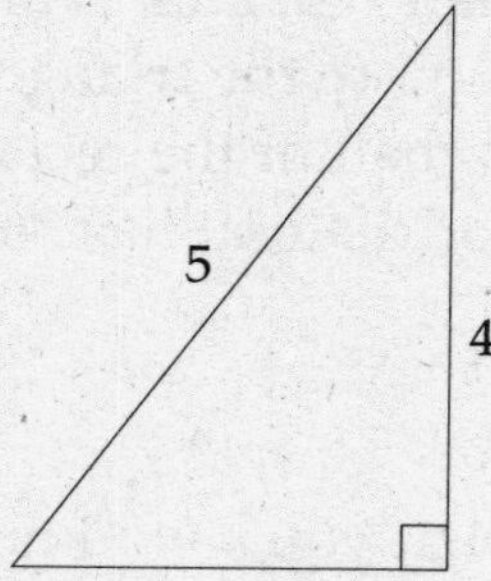

1 If the area of the triangle above is 6, what is its perimeter?

(A) 8 (B) 11 (C) 12 (D) 15 (E) 16

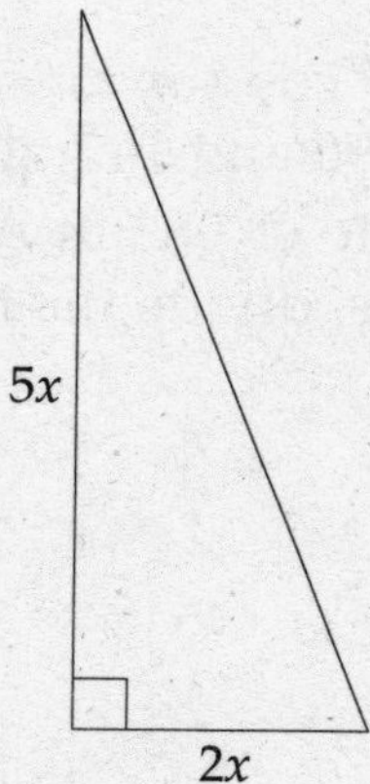

2 If $x = 3$, what is the area of the triangle above?

(A) 10
(B) 12
(C) 21
(D) 30
(E) 45

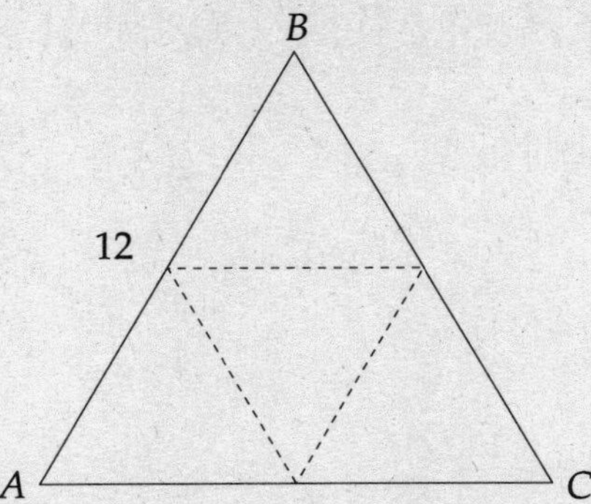

3 If equilateral triangle *ABC* is cut by three lines as shown to form four equilateral triangles of equal area, what is the length of a side of one of the smaller triangles?

(A) 3 (B) 4 (C) 5 (D) 6 (E) 8

MEDIUM

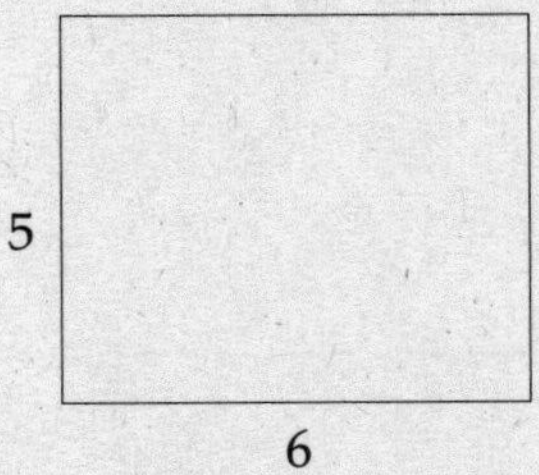

4 If the rectangle above is divided into 2 triangles, then the sum of the perimeters of <u>both</u> triangles is

(A) equal to 30
(B) less than 30
(C) equal to 32
(D) equal to 34
(E) greater than 34

5 A movie theater is 3 blocks due north of a supermarket and a beauty parlor is 4 blocks due east of the movie theater. How many blocks long is the street that runs directly from the supermarket to the beauty parlor?

(A) 2.5 (B) 3 (C) 4 (D) 5 (E) 7

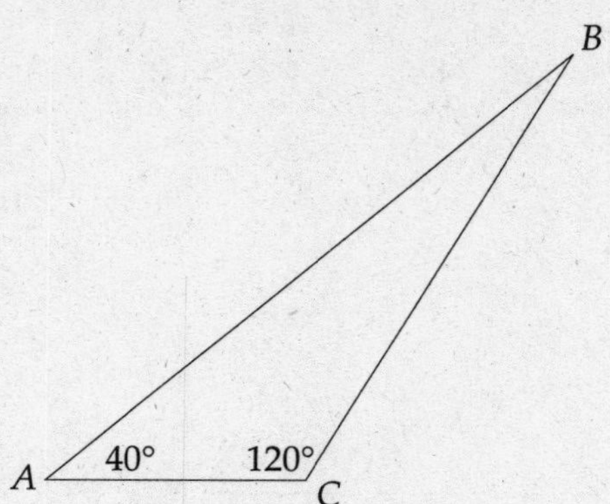

6 In triangle ABC above, if line CD (not shown) bisects AB, which of the following must be true?

I. $DB > BC$
II. $AD = DB$
III. $AB > BC$

(A) I only
(B) II only
(C) III only
(D) I and II only
(E) II and III only

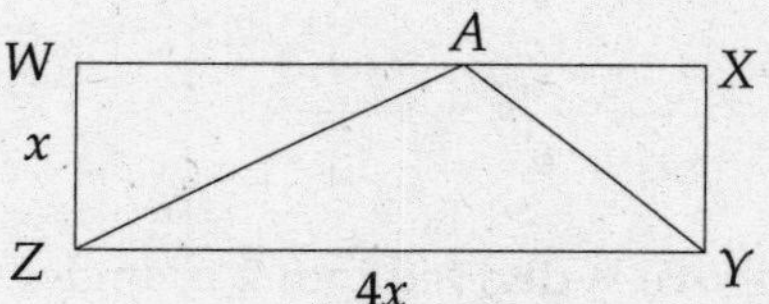

7 What is the area of triangle YAZ?

(A) $3x$ (B) x^2 (C) $5x$ (D) $2x^2$ (E) $4x^2$

HARD

8 If a triangle has vertices of (–1, 5), (–1, –3), and (5, –3), then the perimeter of the triangle is

(A) 8 (B) 10 (C) 15 (D) 24 (E) 30

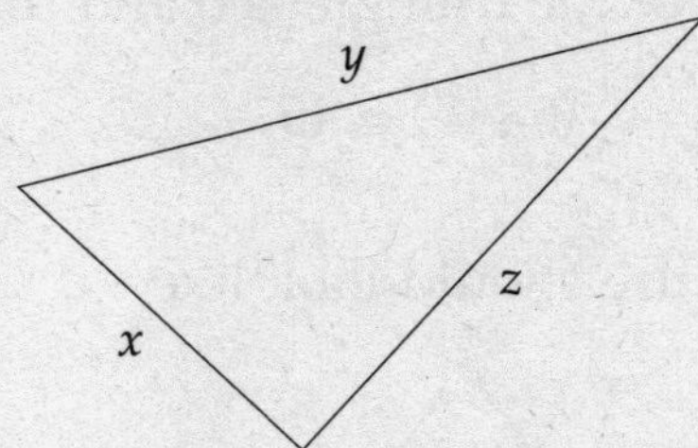

9 If $x = 7$ and $y = 11$ then the difference between the greatest and least possible integer values of z is

(A) 11 (B) 12 (C) 13 (D) 14 (E) 15

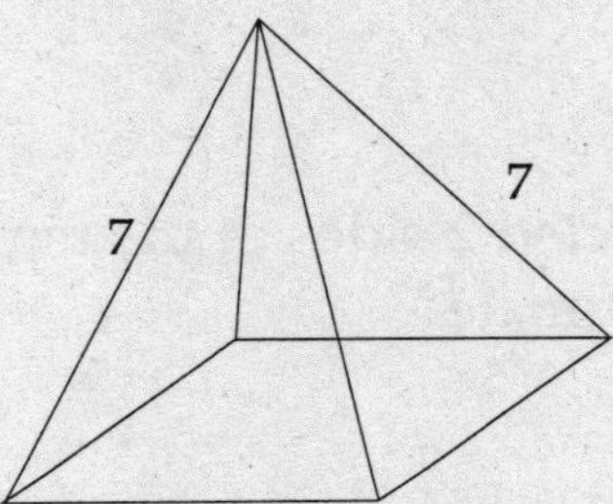

10 The pyramid above has a square base with an area of 16. If the top of the pyramid is cut off to make a smaller pyramid whose base has an area of 4, then the length of the edge from the vertex to the new base is

(A) $\frac{7}{4}$ (B) 2 (C) $\frac{7}{2}$ (D) 4 (E) 5

Answers and Explanations: Problem Set 15

EASY

1 C Several ways to get this question: you could recognize that it's a Pythagorean triple (3:4:5), which would give you the length of the unmarked leg. Or you could set up the following equation:

$$a=\frac{1}{2}bh$$

$\frac{1}{2}b(4) = 6$, so $2b = 6$ and $b = 3$. All we did was substitute the height and the area, which are both given in the problem, into the formula for the area of a triangle.

2 E If $x = 3$, then the base of the triangle is 6 and the height is 15. That would mean the area $= \frac{1}{2}(6)(15) = 45$.

3 D Here's what your picture should look like:

Each of our 3 lines bisected 2 sides of the big triangle, so each side is 6. (Don't forget to estimate.)

MEDIUM

4 E First draw a diagonal, which divides the rectangle into 2 triangles. Now estimate: how long does that diagonal look? It's definitely longer than 6—so the perimeter of each triangle must be bigger than 17, so both perimeters must be greater than 34. Cross out A, B, and C. Now be more precise. The diagonal looks like it's about 8, wouldn't you say? So the perimeter of 1 triangle is around 19, and the perimeter of both is around 38. Pick D. (To figure this exactly, you have to use the Pythagorean Theorem: $5^2 + 6^2 = c^2$. So $c^2 = \sqrt{61}$, which is a little less than 8.)

5 D Draw a little map, which should look like this:

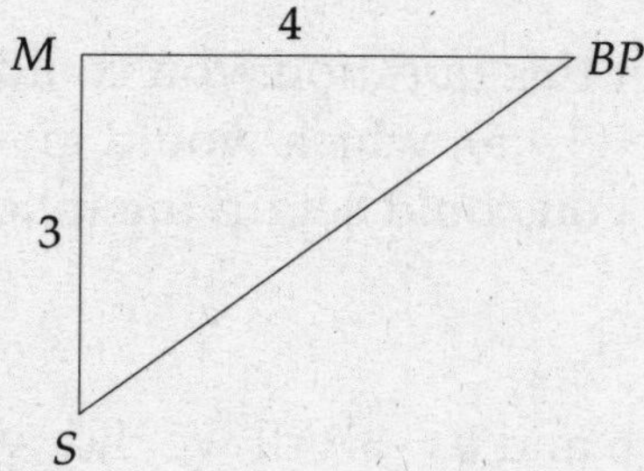

Now you have a 3:4:5 right triangle, so the street from the supermarket to the beauty parlor is 5 blocks long.

6 **E** Draw a line from C that bisects AB. It should look like this:

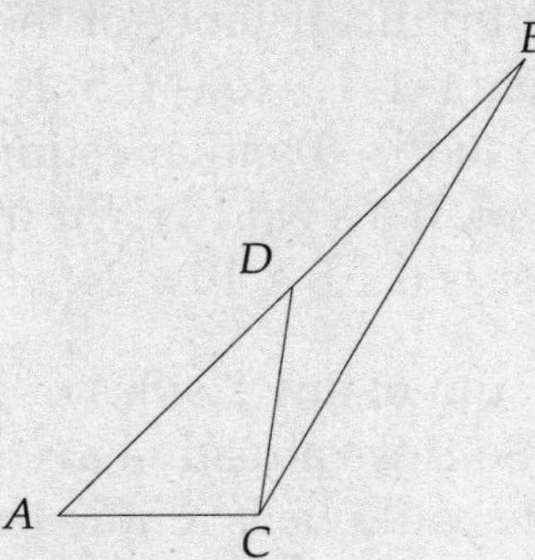

Now you can estimate. DB isn't longer than BC, is it? Cross out A and D. Is AD longer than DC? Yes, D is the midpoint of AB, remember?—cross out C. Is AB longer than BC? Definitely, so the answer is E. The rule for triangles is that the longest leg is opposite the widest angle; the shortest leg is opposite the smallest angle. Draw yourself a few triangles and you'll see it works.)

HARD

7 **D** Let's plug in. If $x = 2$, then ZY is 8 and WZ is 2. (Write that on your diagram.) To get the area of YAZ, notice that WZ is the height of the triangle, so $\frac{1}{2}(8)(2) = 8$. Plug 2 back into the answer choices, and D is $2(2^2) = 8$.

8 **D** Make yourself a grid so you can see what you're doing. It should look like this:

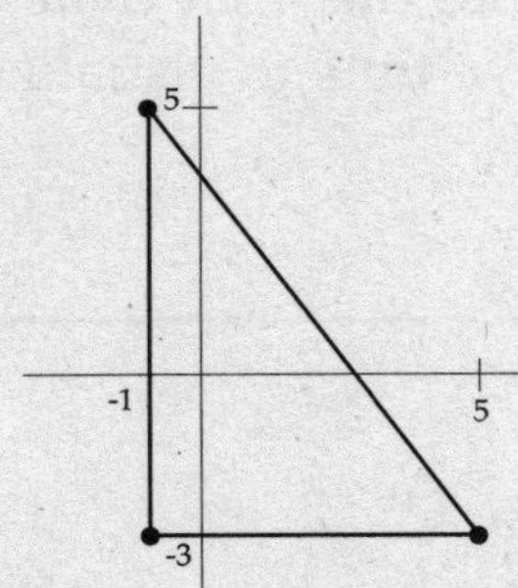

Obviously your drawing won't be exactly to scale but don't worry about it. Now count up the distance of the sides. From (–1, –3) to (5, –3) is 6 units. From (–1, –3) to (–1, 5) is 8 units. To figure out the distance from (–1, 5) to (5, –3) either estimate or use the Pythagorean Theorem. It's a 6:8:10 right triangle, so that distance is 10, and the perimeter is 6 + 8 + 10 = 24.

9 **B** Here's the rule: the sum of any 2 sides of a triangle must be more than the third side. So if we already have sides of 7 and 11, the longest the third side could be is a little less than 18. Since the third side has to be an integer, the longest it could be 17. Now for the shortest possible length of the third side: 11 – 7 = 4, so the third side has to be an integer bigger than 4—that's 5. So the difference between the greatest possible and the least possible is 17 – 5 = 12.

10 **C** If the area of the base is 16, the side of the base is 4. (Don't forget that the base is square.) If we want our new pyramid to have a base with an area of 4, then the new side will be 2. Any face of the original pyramid will be similar to any face of the new pyramid because all the angles stay the same. Since the triangles are similar, the legs are in proportion. If the base goes from 4 to 2, we cut it in half, so the side goes from 7 to $\frac{7}{2}$. Or you could set up a proportion: $\frac{4}{7} = \frac{2}{x}$, with $\frac{4}{7}$ as the original pyramid's base-to-edge, and $\frac{2}{x}$ as the smaller pyramid's base-to-edge.

TIP: On problems like this that deal with complex shapes, give yourself plenty of time to visualize the problem before trying to solve it.

PROBLEM SET 16: CIRCLES

EASY

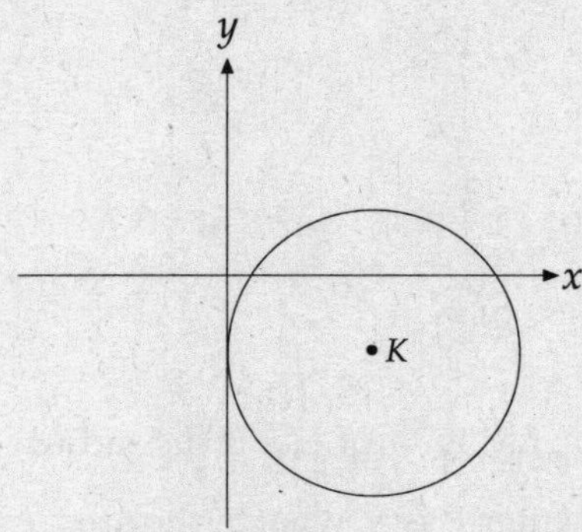

1 Point K is the center of the circle above, and the coordinates of Point K are (2, –1). What is the area of the circle?

(A) π (B) 2π (C) 4π (D) 6π (E) 8π

2 Circle P has a radius of 7 and Circle R has a diameter of 8. The circumference of Circle P is how much greater than the circumference of Circle R?

(A) π (B) 6π (C) 8π (D) 16π (E) 33π

3 Eight circular tiles are arranged as above as part of a pattern in a floor design. Each tile has a diameter of 1 inch. If the group of 8 tiles is enclosed by a diamond-shaped border that lies tangent to the exterior tiles, what is the total length, in inches, of the border?

(A) 4 (B) 6 (C) 8 (D) 9 (E) 12

MEDIUM

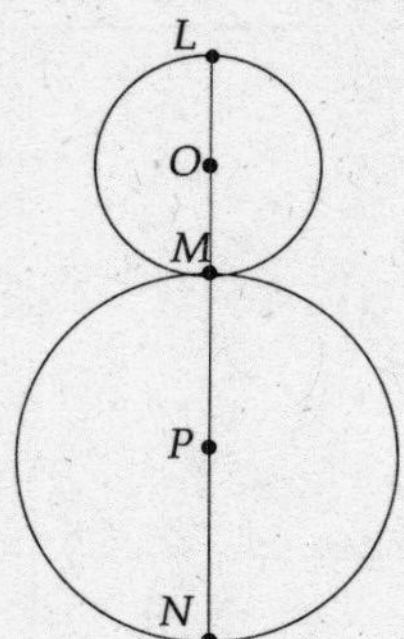

4 In the figure above, LM is $\frac{1}{3}$ of LN. If the radius of the circle with center P is 6, what is the area of the circle with center O?

(A) 4π
(B) 9π
(C) 12π
(D) 18π
(E) 36π

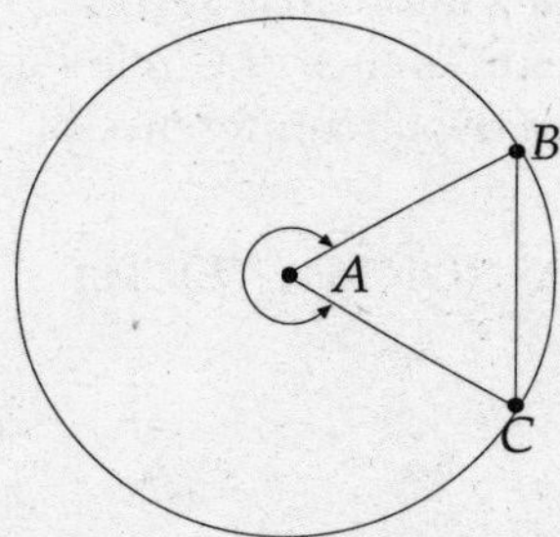

Circle X

5 In the figure above, *Circle X* has center A, and $BC = AB$. What is the degree measure of the marked angle?

(A) 60°
(B) 180°
(C) 270°
(D) 300°
(E) 340°

6 What is the greatest number of distinct regions that could be formed by a circle overlapped by a triangle?

(A) 3 (B) 4 (C) 6 (D) 7 (E) 8

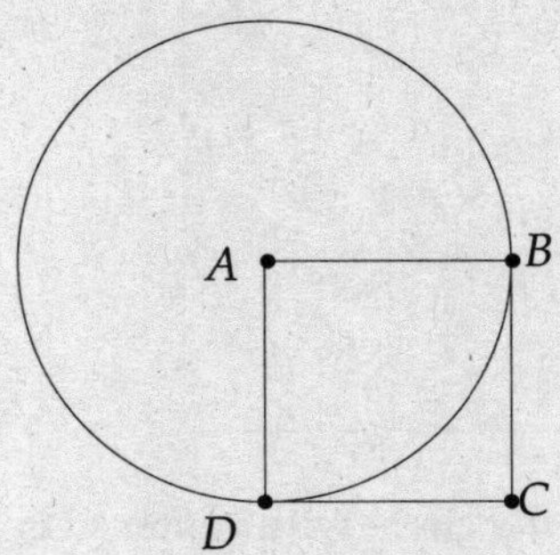

7 Points D and B lie on the circle above with center A. If square $ABCD$ has an area of 16, what is the length of arc BD?

(A) 2π (B) 4 (C) 8 (D) 4π (E) 8π

HARD

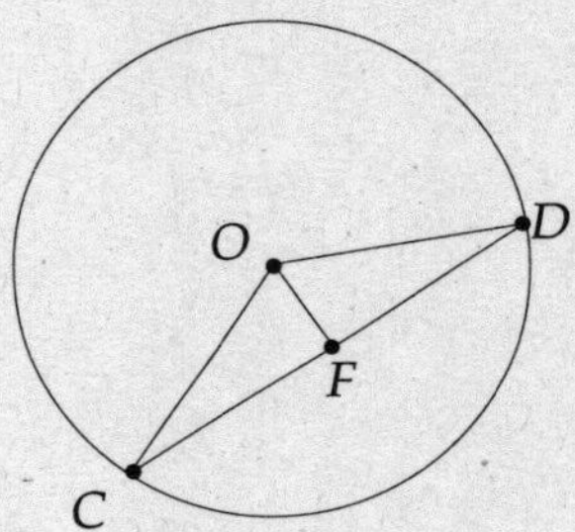

8 In the figure above, what is the circumference of the circle if COD is 120° and OF bisects CD and with center OF has a length of 1.5?

(A) $\frac{2\pi}{3}$

(B) $\frac{3\pi}{2}$

(C) 3π

(D) 6π

(E) 9π

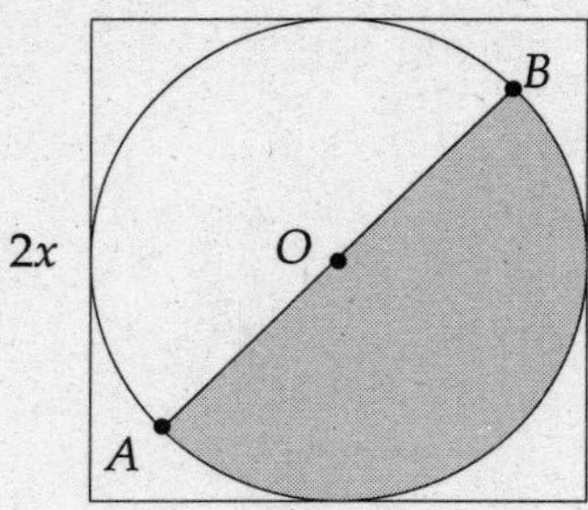

9 In the figure above, the circle with center O is inscribed in a square with sides of $2x$. If AB is a diameter of the circle, then in terms of x, what is the total area of the unshaded regions?

(A) $4x^2 + \pi x^2$

(B) $4x + \frac{\pi x^2}{2}$

(C) $4x^2 - \frac{\pi x^2}{2}$

(D) $4x^2 - \pi x$

(E) $2x - \frac{\pi x^2}{2}$

10 In a certain machine, a gear makes 12 revolutions per minute. If the circumference of the gear is 3π inches, approximately how many <u>feet</u> will the gear turn in an hour?

(A) 6782
(B) 565
(C) 113
(D) 108
(E) 9

Answers and Explanations: Problem Set 16

EASY

1 **C** Count up the units of the radius—it's 2. Then use the formula for area of a circle: $\pi r^2 = \pi(2^2) = 4\pi$.

2 **B** The circumference of Circle P is $2\pi r = 2\pi(7) = 14\pi$. The circumference of Circle R is $2\pi r = 2\pi(4) = 8\pi$. Now just subtract. If you picked E, you calculated area instead of circumference. Hello?

3 **E** Draw the diamond-shaped border around the circles. Write the diameter of a circle on your diagram too. Each side of the border is 3 diameters, or 3 inches—so all four sides will have $4 \times 3 = 12$ inches.

MEDIUM

4 **B** Write in 6 by the radius of the bigger circle. That makes the diameter of the bigger circle 12. If *LM* (the diameter of the smaller circle) is $\frac{1}{3}$ the length of *LN*, the equation is $\left(\frac{1}{3}\right)(12 + x) = x$. $4 = x - \frac{x}{3}$, and $x = 6$. (You don't have to write an equation. You could estimate and try some numbers. Doesn't *LM* look like it's about half of *MN*? It is.) If the diameter of the smaller circle is 6, then its radius is 3 and its area is 9π.

5 **D** Estimate first. The marked angle is way over 180—in fact, it's not that far from 360. Cross out A and B. You know $AC = AB$ because they're both radii. That means $BC = AB = AC$, and that triangle is equilateral. So angle *BAC* is 60°. Subtract that from 360 and you're in business. (Even if all you could do was estimate, go ahead and take a guess.)

6 **D** Draw a few diagrams and see how many regions you can come up with. Here's what the diagram should like:

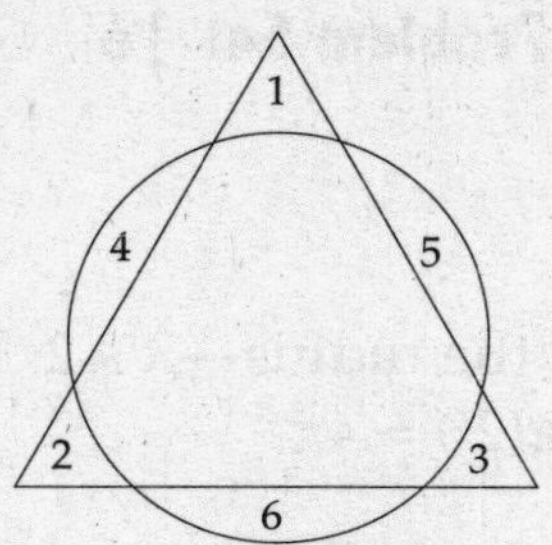

7 **A** If the square has an area of 16, then the side of the square is 4. Write that on your diagram. Now you know the radius of the circle is also 4, so the circumference is 8π. Angle *BAD* has 90°, since it's a corner of the square. And since 90 is $\frac{1}{4}$ of 360, arc *BD* is $\frac{1}{4}$ of the circumference. (Pretty cool, huh?) So arc *BD* is $\left(\frac{1}{4}\right)(8\pi)$, or 2π. If you estimated first, as we hope you did, you could have crossed out D and E, and maybe even C.

HARD

8 **D** Write the info on your diagram. If *OF* bisects *CD*, it also bisects angle *COD*, making two 60° angles. Now there are two 30:60:90 triangles. If the shortest leg of one of those triangles is 1.5, then the hypotenuse is 2×1.5, or 3. Aha! That distance is also the radius of the circle, so the circumference is 6π.

9 **C** This one's pretty ugly. To get the unshaded part of the square, we're going to get the area of the whole square and then subtract the shaded part. If the side of the square is $2x$, then the area is $4x^2$. (Cross out E.) The diameter of the circle is equal to the side of the square, even if the diagram shows it tipped to one side. The radius, then, is half the side of the square, or x. The area of the whole circle is πx^2, and the area of the shaded semicircle is $\frac{\pi x^2}{2}$. Now all we have to do is subtract that shaded area from the area of the square, and we get $4x^2 - \frac{\pi x^2}{2}$.

If all those x's made you nervous, you could of course plug in something simple for x, like 2. It helps to plug in on geometry, because it's easier to think of lengths as concrete numbers rather than as variables.

10 **B** First make $\pi = 3$, so the gear travels 9 inches per revolution. If it makes 12 revolutions per minute, that's $12 \times 9 = 108$ inches per minute. Which is $108 \times 60 = 6480$ inches in an hour. To calculate feet, simply divide by 12, and you get 540. Did you pick *A*? They asked for feet, my friend, not inches. Read carefully.

TIP: One more thing—circle questions tend to appear most often in the late-medium and hard questions.

PROBLEM SET 17: QUADRILATERALS, BOXES, AND CANS

EASY

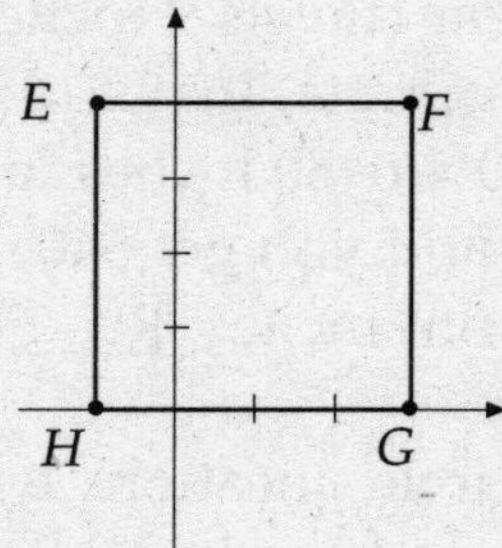

1 The coordinates of Point E are $(-1, 4)$ and the coordinates of Point H are $(-1, 0)$. If $EFGH$ is a square, what are the coordinates of Point F?

(A) (3, 4)
(B) (3, 0)
(C) (3, 5)
(D) (4, 1)
(E) (4, –1)

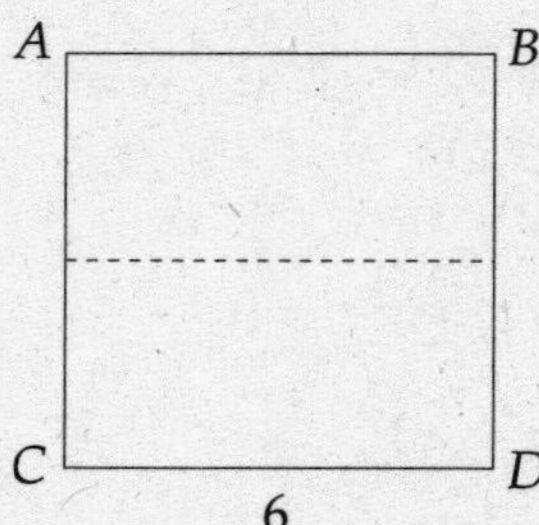

2 The square above is folded on the dotted line so that A is directly on top of C. If the square is folded again so that B is on top of C, what is the length of the side of the new square?

(A) 2
(B) 3
(C) $3\sqrt{2}$
(D) 4
(E) $3\sqrt{3}$

3 How many squares with sides of 1 could fit into the rectangle above?

(A) 3 (B) 4 (C) 6 (D) 9 (E) 12

MEDIUM

4 The area of Rectangle *K* is three times the area of Rectangle *Q*. The area of rectangle *Q* is twice the area of Rectangle *P*. If the area of Rectangle *Q* is 4, what is the difference between the area of Rectangle *K* and Rectangle *P*?

(A) 12 (B) 10 (C) 8 (D) 6 (E) 2

5 Rectangle *LMNO* is divided into 8 square sections. If the area of *LMNO* is 24, what is the total area of the shaded regions?

(A) 12 (B) 9 (C) 8 (D) 6 (E) 2

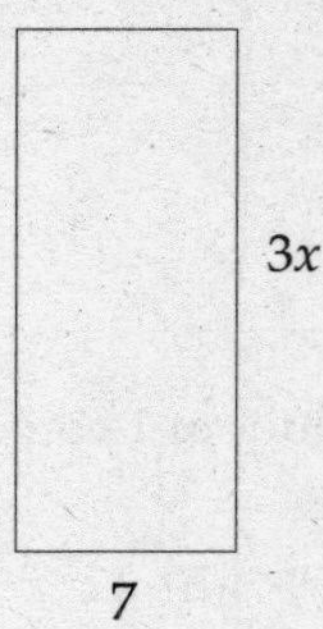

Note: Figure not drawn to scale.

6 If the area of the rectangle above is 42, what is the value of x^2?

(A) 24
(B) 21
(C) 6
(D) 4
(E) 2

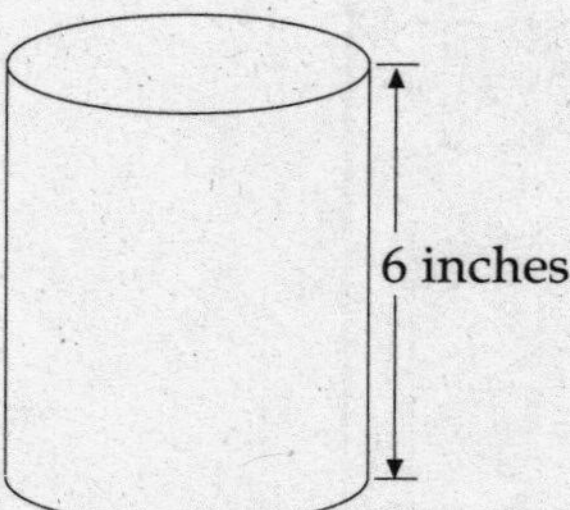

7 In the figure above, the radius of the base of the cylinder is its height. What is the volume of the cylinder in cubic inches?

(A) 9π
(B) 15π
(C) 18π
(D) 36π
(E) 54π

HARD

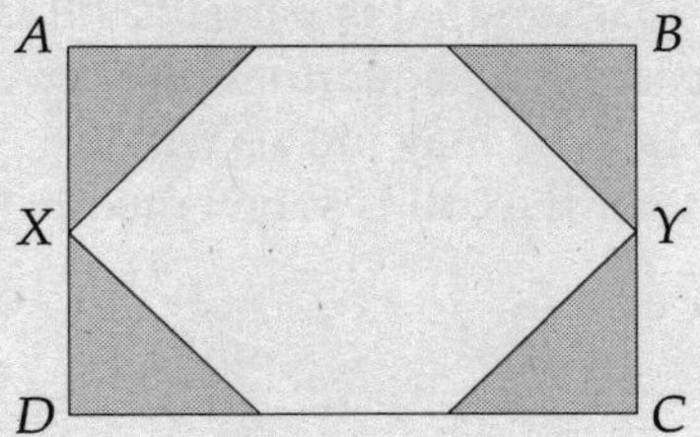

8 The area of rectangle $ABCD$ is 96, and $AD = \frac{2}{3}(AB)$. Points X and Y are midpoints of AD and BC, respectively. If the 4 shaded triangles are isosceles, what is the perimeter of the unshaded hexagon?

(A) 16
(B) $8 + 6\sqrt{2}$
(C) 24
(D) $8 + 16\sqrt{2}$
(E) $16 + 24\sqrt{2}$

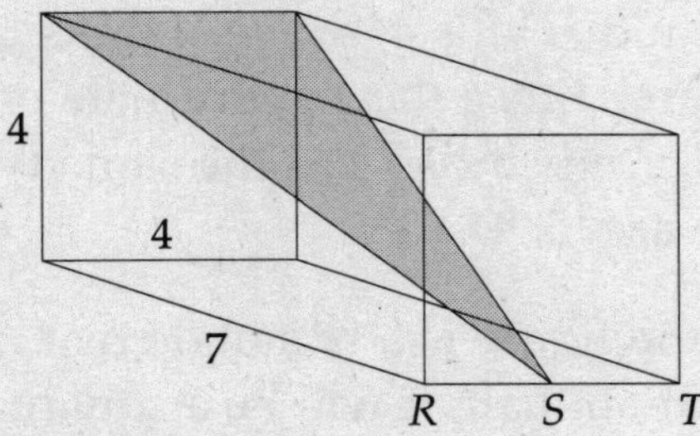

9 In the figure above, S is the midpoint of RT. What is the area of the shaded triangle?

(A) 14
(B) 16
(C) $2\sqrt{65}$
(D) 18
(E) $4\sqrt{6}$

10 A ball with a volume of 18 cubic inches is dropped into an aquarium that is partially filled with water. If the base of the aquarium measures 12 inches by 6 inches, how may inches will the level of water rise after the ball is submerged?

(A) $\frac{1}{4}$

(B) $\frac{1}{2}$

(C) 1

(D) 4

(E) 6

Answers and Explanations: Problem Set 17

EASY

1 **A** Write the coordinates of E and H on your diagram. What's the distance between E and H? It's 4, so F is 4 units to the right of E. (The figure is a square.) Since the x-coordinate of E is -1, four units to the right of E is 3. And since F has the same y-coordinate as E, the coordinates of F are (3, 4).

2 **B** After the first fold, the figure is a rectangle that measures 6 by 3. If it is then folded again, in half, it will be a square again, with sides of 3. This is a visual perception question—if you had trouble with it, cut a square piece of paper and fold it twice, following the directions in the question, and see what happens.

3 **E** Drawing on the diagram could help. How many sides of 1 can fit along the long edge of the rectangle? 6. And how many rows will fit along the short edge? 2. Now just multiply $6 \times 2 = 12$. Or draw them in and count them up.

MEDIUM

4 **B** If the area of Q is 4, then the area of P is 2 and the area of K is 12. So the difference between the areas of K and P is $12 - 2 = 10$. Don't bother writing equations to solve this problem—it's a waste of time. Just use the given number (4), and figure out the other areas one at a time. Or just look at the diagram and see that a total of 3 out of the 8 squares are shaded: $\frac{3}{8} \bullet \frac{24}{1} = 9$.

5 **B** If the total area is 24, and there are 8 sections, then each section has an area of 3. The six triangles are half-sections, so each triangle has an area of 1.5. And $1.5 \times 6 = 9$. If you estimate, you should be able to cross out E, at least.

6 **D** $7 \times 3x = 42$, so $21x = 42$ and $x = 2$. But don't pick E! One last thing—the questions asks for x^2, so $2^2 = 4$. (A lot of SAT questions are testing your ability to follow directions, so make sure you reread the question carefully before bubbling in your final answer.)

7 **E** If the radius is half the height, then the radius is 3. To get the volume of a cylinder, multiply the area of the base times the height—in this case, $\pi(3^2) \times 6 = 54\pi$.

HARD

8 **D** First calculate the dimensions of the rectangle. If the area is 96, and the width is $\frac{2}{3}$ of the height, your equation is $x \bullet \left(\frac{2}{3}\right)x = 96$. $\frac{2x^2}{3} = 96$, $x^2 = 144$, and $x = 12$. So the long side is 12 and the short side is $\frac{2}{3}$ of 12, or 8. If X and Y are midpoints, then AX, XD, BY, and YC all equal 4. Since the triangles are isosceles, the legs of the triangles (not the hypotenuses) are also 4. Here's what we have so far:

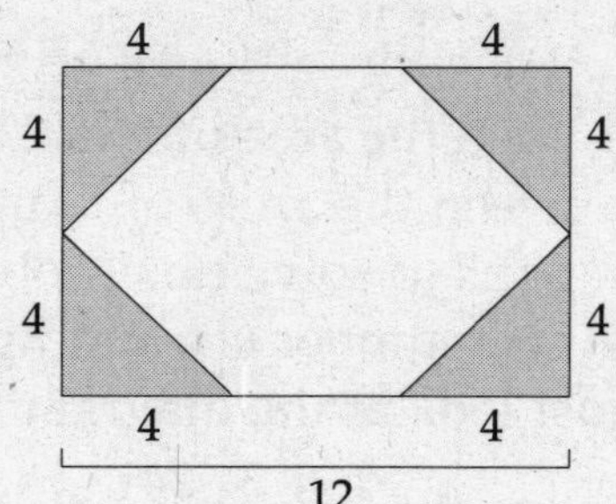

The unmarked piece of the long side of the rectangle is also 4, since the side is 12 altogether. Since we have isosceles *right* triangles, the hypotenuses are $4\sqrt{2}$. Now add it all up: the 4 hypotenuses would be $4(4\sqrt{2}) = 16\sqrt{2}$, and the sides of the hexagon that lie along the rectangle would be $2(4) = 8$. So the total perimeter of the hexagon is $8 + 16\sqrt{2}$.

9 **C** Part of this problem is easy—the base of the triangle is 4. Now for the hard part: imagine drawing an altitude of the triangle from the center of the base all the way to the tip. That's the same as drawing a diagonal of the side of the face measuring 4 by 7. So to calculate the altitude of the triangle, use the Pythagorean Theorem to find the hypotenuse of the right triangle formed by drawing a diagonal on the side of the box. $4^2 + 7^2 = c^2$, $16 + 49 = c^2$, $c = \sqrt{65}$. That means the area of the shaded triangle is $\frac{1}{2}(4)(\sqrt{65}) = 2\sqrt{65}$.

10 **A** The important concept here is that the base of the aquarium remains constant, and the height of the water is *not* constant. (The problem doesn't tell you the height of the aquarium, because it doesn't matter.) If the base is 12 by 6, then its area is 72. If the ball has a volume of 18, then it will make the water rise $\frac{1}{4}$ inch, because $72 \times \frac{1}{4} = 18$.

To clarify: let's say the water in the aquarium was 1 inch deep. Then the volume of water would be $l \times w \times h = 12 \times 6 \times 1 = 72$. If we drop in a ball with a volume of 18, then the new volume of the water would be 72 (new height) = 72 + 18, and the new height = $1\frac{1}{4}$. So the water level went from 1 to $1\frac{1}{4}$ inches deep.

PROBLEM SET 18: MIXED BAG

EASY

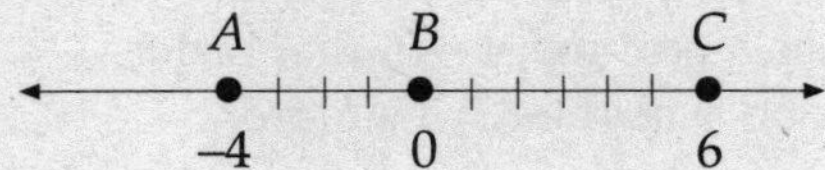

1 On the number line above, what is $BC - AB$?

(A) 0
(B) 2
(C) 4
(D) 6
(E) 10

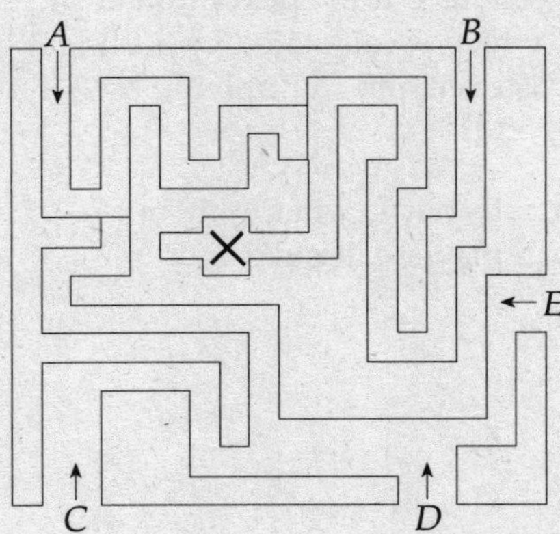

2 Which path allows access to X without crossing any lines?

(A) *A*
(B) *B*
(C) *C*
(D) *D*
(E) *E*

MEDIUM

3 For a desktop, a cabinetmaker designed a border that consisted of a row of inlaid squares of different kinds of wood. If the border started with a square of oak, and continued with cherry, walnut, ash, and maple, in that order, what kind of wood was the 83rd square?

(A) oak
(B) cherry
(C) walnut
(D) ash
(E) maple

4 A certain game is played by the following rules:

1. The first player calls out a number from 1 to 10.
2. The next player calls out the next number in the sequence, which is calculated by adding 3 if the number is even and multiplying by 2 if the number is odd.

If the first player calls out 1, what number should the seventh player call out?

(A) 6
(B) 13
(C) 26
(D) 29
(E) 30

$$\begin{array}{r} AB \\ \times\ BA \\ \hline 4A \\ AB \\ \hline A\,5A \end{array}$$

5 Given the multiplication problem above, in which A and B represent different digits, what is the value of B?

(A) 1
(B) 2
(C) 3
(D) 4
(E) 8

Number of Households	Number of Dogs
1	0
5	1
2	2

Number of Households	Number of Cats
3	0
3	1
2	2

6 The tables above shows the number of cats and dogs owned by the same 8 households. If the households that own either no dogs or 2 dogs also have no cats, then 2 households own

(A) no cats and no dog
(B) no cats and 1 dog
(C) 1 cat and no dogs
(D) 2 cats and 1 dog
(E) 2 cats and 2 dogs

HARD

7 How many distinct numbers between 1 and 40 contain the digit 2, the digit 3, or the digits 2 and 3?

(A) 12
(B) 16
(C) 20
(D) 24
(E) 26

8 Bob has a pile of poker chips that he wants to arrange in even stacks. If he stacks them in piles of 10, he has 4 chips left over. If he stacks them in piles of 8, he has 2 chips left over. If Bob finally decides to stack the chips in only 2 stacks, how many chips could be in each stack?

(A) 14
(B) 17
(C) 18
(D) 24
(E) 34

9 Conchata began a hike at an average speed of 3 miles per hour. Mariel started the same hike 2 hours later. If Mariel hiked at an average speed of 4 miles per hour, how many hours did it take Mariel to catch up to Conchata?

(A) 1
(B) 4
(C) 6
(D) 10
(E) 12

10 How many people must be in a group in order to be certain that at least 2 people in the group have first names that begin with the same letter?

(A) 20 (B) 26 (C) 27 (D) 35 (E) 64

Answers and Explanations: Problem Set 18

EASY

1 **B** BC has a length of 6. AB has a length of 4. So $BC - AB = 2$.

TIP: On number line problems, sometimes you want the distance between 2 points, as in the problem above. And sometimes you want the number of a point on the line—for instance, using this number line, $A + B = -4$, because $A = -4$ and $B = 0$. You can have a negative value for a point on the line, but not a negative distance. Read the problem carefully and mark up your diagram so you don't confuse the two.

2 **B** Kinda fun, isn't it? Definitely use your pencil to trace out the path. In this case A dead-ends at B; and C, D, and E all dead-end at A.

TIP: These maze, map, or find-the-number-of-routes questions show up in the easy, medium, and hard sections. One possible pitfall is that you spend too much time on it, tracing and retracing the path to check your answer, and coming up with different answers. It's only one question—so take your best shot and keep moving.

MEDIUM

3 C This is a pattern question—and you do them all the same way. How many elements are in the pattern? 5, in this case. Divide 5 into the total, which is 83. You get 16 with a remainder of 3. That means the whole pattern is repeated 16 times and continues 3 elements past that. Just count to the third element in the pattern, which is walnut.

TIP: In a pattern question, don't convert the remainder of the division problem into a fraction. In other words, if the pattern has 3 elements and the total is 7, your dividend should be 2 remainder 1, not $2\frac{1}{3}$.

4 D This is a "game" problem, and you solve it the same way you solve function problems—just follow the directions. The rules are: add 3 if the number is even; multiply by 2 if the number is odd. Since our first number is 1, the sequence should go 1, 2, 5, 10, 13, 26, 29. Make sure you go all the way to the 7th player.

TIP: Don't calculate the steps in a game question in your head. Even if all you're doing is adding 3 and multiplying by 2, write down each number in the sequence. It's too easy to make a careless mistake otherwise—and nobody's giving you extra points for not marking up your test.

5 A Yeah, we know, these problems are a drag. How do you start? First take a quick look at the problem to see if you can draw any conclusions about what the variables represent. In the addition part of the problem you see that 4 + B is 5, so B must be 1. That's it. If the question was harder, it might ask for A. So let's figure that at two. Rewrite the problem to one side, filling in 1 for all the B's. Now for A: the question says they are distinct, so A can't be 1. Try 2. That gives you $21 \times 12 = 252$. It works.

TIP: These questions are a bummer because sometimes it's hard to know where to begin. If you don't see any starting point, just try some numbers and see what happens. If you try a couple of different things and nothing seems to work, skip the question and come back to it later. These questions appear most often in the easy and medium sections. Depending on how the question is asked, you may be able to backsolve.

6 D Do this one step at a time. There's only one household with no dogs. Draw a line from that household to the row with no cats. Then draw a line from the 2 households with 2 dogs to the households with no cats. We're left with the 5 households that have 1 dog—and on the other table, 3 households have 1 cat and 2 households have 2 cats. So 3 households have 1 cat and 1 dog, and 2 households have 2 cats and 1 dog.

TIP: It's confusing, all these dogs and cats. It would be mighty easy to get the two categories mixed up. Just don't go too fast and start over again if you lose your grip.

HARD

7 D Write them all down, in an orderly fashion, one digit at a time. The numbers containing 2 are: 2, 12, 20, 21, 22, 23, 24, 25, 26, 27, 28, 29, 32. The numbers containing 3 are: 3, 13, 23, 30, 31, 32, 33, 34, 35, 36, 37, 38, 39. Not so fast! This is a hard question, remember? You need to check to see whether any numbers are repeated—and sure enough, 23 and 32 are on both lists. So cross off the duplicates, and *then* count up the numbers.

TIP: On this kind of problem, don't try anything fancy. Just write down the list, being careful not to miss anything, and then count up how many you have. On a hard question, there's likely to be some kind of sneaky trap, as in the question above. Be wary.

8 B This is a remainder question. If the chips are in stacks of 10, there are 4 left over. That means there's a total of 14 chips, or 24, or 34, or any number × 10 with 4 added. If the chips are in stacks of 8, there are 2 left over, which means there are 10, 18, 26, or 34 . . . or any number × 8 with 2 added. 34 works for both 10-chip stacks with remainder 4, and 8-chip stacks with remainder 2. But the question asks for the number of chips in each stack, so you have to divide 34 by 2. You could also backsolve by dividing the answer choices first to get the total number of chips.

9 C It never hurts to make a little diagram:

Hours:	1	2	3	4	5	6	7	8
miles/Conchata:	3	6	9	12	15	18	21	24
miles/Mariel:	0	0	4	8	12	16	20	24

All we're doing is keeping track of their progress, hour by hour. Mariel catches up to Conchata at 24 miles, at which point Conchata had hiked for 8 hours and Mariel for 6. (They must be in pretty good shape.)

TIP: You could solve this question by writing an equation based on the formula *rate* · *time* = *distance*, but it's much harder that way. You have to realize that the distance is equal, and that Conchata's time is Mariel's time plus 2 hours, giving you the equation $3(x + 2) = 4x$. It's hard to come up with that kind of equation when you're under time pressure. We like the chart much better.

10 C In order to be certain, you have to allow for the worst case scenario—every person's name begins with a different letter. So the first person's name begins with A, the second with B, the third with C, etc. etc. When you get all the way to Z, you have 26 people, so the 27th person's name will have to begin with a letter you already have, no matter what the letter is.

TIP: Don't pick anything obvious. A lot of people are going to pick 26 for this question, because there are 26 letters in the alphabet. On a hard question, there's got to be more to it than that.

PROBLEM SET 19: GRID-INS

EASY

16 If $x - y = -6$, then y is how much greater than x?

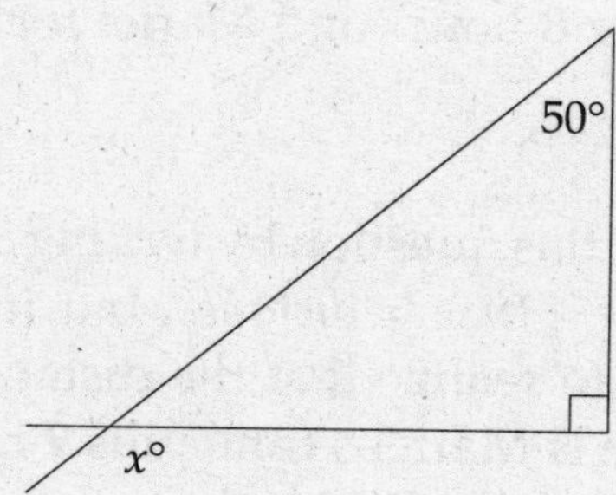

17 What is the value of x?

	Day 1	Day 2	Day 3
pies	15	12	10
cakes	6		
toys	16		15
hats	5	22	

18 The unfinished chart above shows the sales for a certain booth at a three-day fair. If, at the end of the fair, the owners of the booth sold more pies than toys, then how many toys could have been sold on Day 2?

MEDIUM

19 If $\{x\}$ is defined as the number of distinct prime factors of x, what is the value of $\{15\} - \{38\}$?

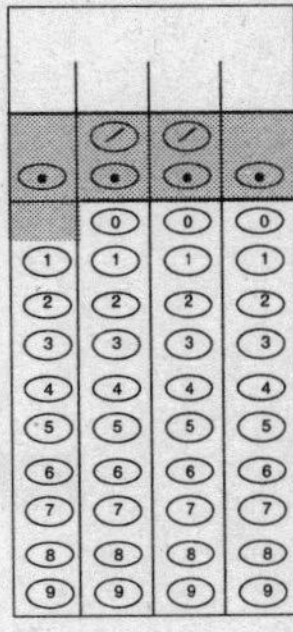

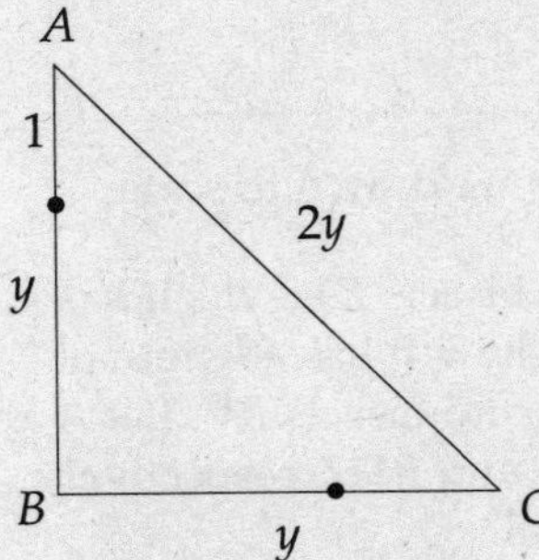

20 If $y = 5$, what is the length of BC?

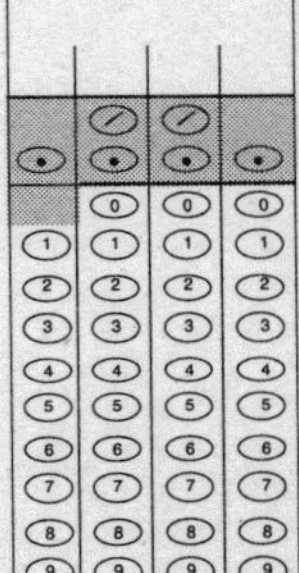

21 If x is an integer and $\frac{x}{4}$ is less than .8 and greater than .25, what is one possible value of x?

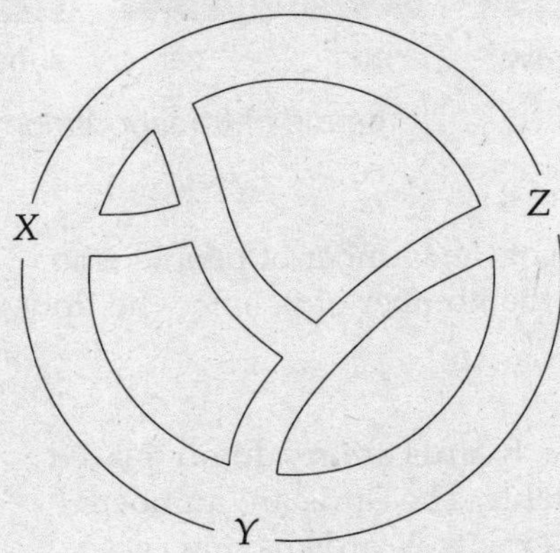

22 On the map above, X represents a theater, Y represents Chris's house, and Z represents Peter's house. Chris walks from his house to Peter's house without passing the theater, and then walks with Peter to the theater without walking by his own house again. How many different routes can Chris take?

HARD

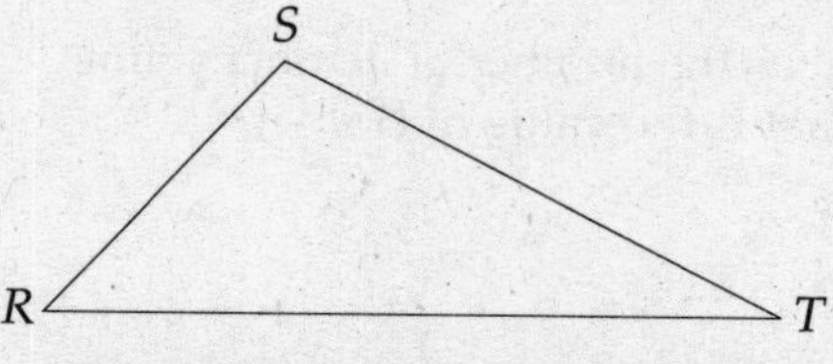

Note: Figure not drawn to scale.

23 Line SV (not shown) bisects RT, which has a length of 8. SU (not shown) has a length of 7, is perpendicular to RT, and bisects RV. Let A and a be the areas of SVT and RSU, respectively. What does $A - a$ equal?

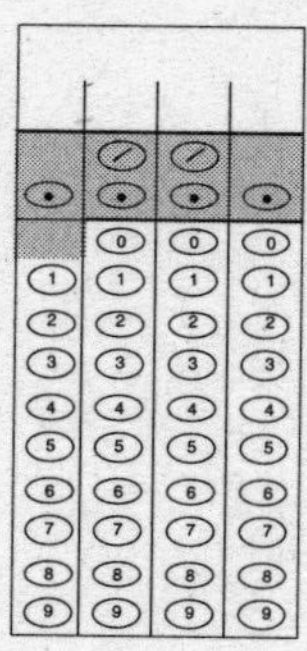

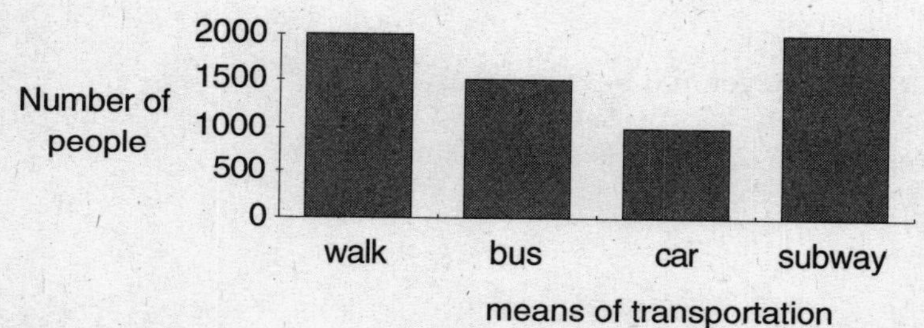

24 What is the ratio of the number of people who walk to work to the number of people who do not walk to work?

25 In a certain game, 8 cards are randomly placed face-down on a table. The cards are numbered from 1 to 4 with exactly 2 cards having each number. If a player turns over two of the cards, what is the probability that the cards will have the same number?

Answers and Explanations: Problem Set 19

EASY

16 **6**

Set the equation equal to y, since that's what the question asks for. You get $-y = -6 - x$. Multiply through by -1 and you get $y = 6 + x$, so the answer is 6. For an easier solution, you could also plug in here: say $x = 2$ and $y = 8$, which satisfies the equation. Then y is equal to x plus 6.

17 **140**

The unmarked angle in the triangle is 40°, since triangles have 180° and the other angles are 50 and 90. The 40° angle and x lie on a straight line, so $40 + x = 180$, and $x = 140$.

18 **0, 1, 2, 3, 4, or 5**

The total number of pies sold was 15 + 12 + 10 = 37, and so far the number of toys sold was 16 + 15 = 31. If they sold more pies than toys, they could sell anything from 0 to 5 toys and still be under 37. It doesn't matter which number you grid in.

MEDIUM

19 **0**

The distinct prime factors of 15 are 3 and 5, so {15} is 2. (The function definition is the number of distinct prime factors. It doesn't matter what the factors are, just how many of them there are.) 38 has 2 distinct prime factors as well: 2 and 19. So {38} is 2. That means {15} – {38} = 2 – 2 = 0.

20 **3**

As always, write the info on your diagram: if $y = 5$, then $AB = 6$ and $AC = 10$. Use the Pythagorean Theorem to figure out BC, or notice that it's a 6:8:10 right triangle—either way, $BC = 8$.

21 **3 or 2**

Deal with the "less than" and "greater than" parts of this problem one at a time. Convert .8 to a fraction: it's $\frac{8}{10}$. Now try plugging in some numbers for x: if $x = 1$, $\frac{1}{4}$ is less than $\frac{8}{10} \bullet \frac{2}{4}$ is less than $\frac{8}{10} \bullet \frac{3}{4}$ is less than $\frac{8}{10}$. That's it—so far, x could be 1, 2, or 3. Now for the other part of the problem. Convert .25 to a fraction: it's $\frac{1}{4} \bullet \frac{x}{4}$ has to be greater than $\frac{1}{4}$, so x can't be 1. But it could be either 2 or 3.

22 **6**

Here they are:

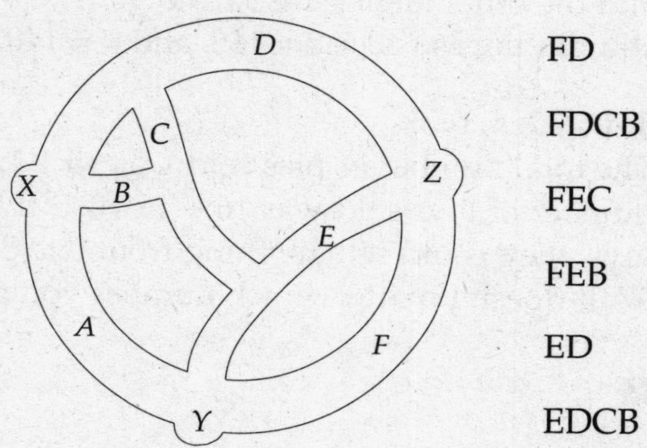

FD

FDCB

FEC

FEB

ED

EDCB

(You don't need to write in the extra letters—we just did it so we could show all the routes.)

This kind of question could keep you going forever, especially since it isn't even multiple choice. Do it a couple of times, grid in your best guess, and keep going. It's impossible to feel very secure about an answer to this type of question because you always feel like you're overlooking something. Don't worry about it—that's just the nature of the question.

HARD

23 **7**

Draw the 2 new lines and write in the given lengths. Your diagram should look like this:

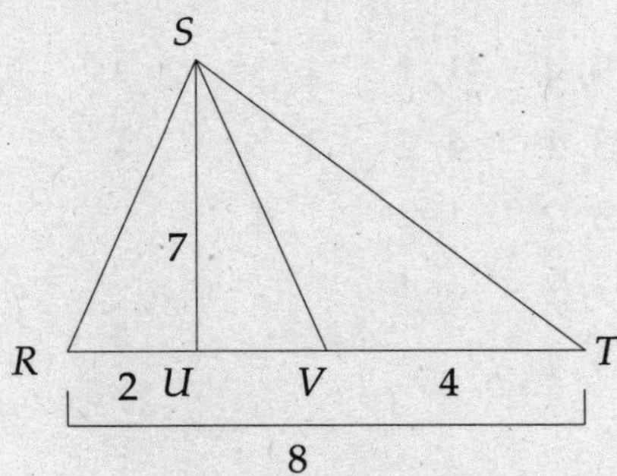

The altitude of both triangles is 7. So if the base of *SVT* is 4 and the base of *RSU* is 2, $A = 14$ and $a = 7$. And as you most certainly know, $14 - 7 = 7$. The whole trick of this question is reading carefully and following the directions so your diagram is correct.

24 $\frac{4}{9}$ **or .44 or .444.**

The number of people who walk to work is 2000. The number of people who do NOT walk to work is 1500 (bus) + 1000 (car) + 2000 (subway) = 4500. The ratio then is $\frac{2000}{4500}$, which you could simply divide on your calculator or reduce, so the number will fit on the grid.

25 $\frac{8}{56}$, $\frac{4}{28}$, **or** $\frac{1}{7}$.

This is a little tricky: the first card doesn't matter, because the problem doesn't ask you to match a particular number, just any number. So the probability of picking any card is 8 out of 8, or $\frac{8}{8}$. Now for the second card, you want to match whatever you drew the first time. There are 7 cards left, and only 1 of them will match your first card, so there's a 1 out of 7 chance of picking it.

$\frac{8}{8} \bullet \frac{1}{7} = \frac{8}{56}$, or $\frac{1}{7}$.

The other way to do this question is to write out all the possibilities. The cards are numbered 1, 2, 3, 4 and 1, 2, 3, 4. Taking each card one at a time, you can make a list of the 2 cards the player turns over:

1, 2	2, 3	3, 4	4, 1	1, 2	2, 3	3, 4
1, 3	2, 4	3, 1	4, 2	1, 3	2, 4	
1, 4	2, 1	3, 2	4, 3	1, 4		
1, 1	2, 2	3, 3	4, 4			
1, 2	2, 3	3, 4				
1, 3	2, 4					
1, 4						

Now count up the number of matches: 4. The total is 28. $\frac{4}{28}$ is $\frac{1}{7}$.

Does that seem like it would take too long? It doesn't really, once you get going. Write them out in order, and don't backtrack.

PROBLEM SET 20: MORE GRID-INS

EASY

16 If $2x - 3y = 7$ and $y = 3$, then what is the value of x?

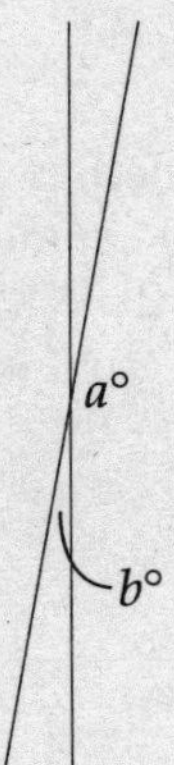

17 In the figure above, if $a = 170$, what is the value of b?

18 At a certain beach, the cost of renting a beach umbrella is $4.25 per day or $28.00 per week. If Kelly and Brandon rent a beach umbrella for 2 weeks instead of renting one each day for 14 days, how much money, in dollars, will they save?

MEDIUM

19 If $x^3 = 27$, then what does 5^x equal?

20 The "zip" of a number is defined as any positive integer raised to a power of the same number. For example, the "zip" of 2 is 2^2. What is the greatest possible value of the "zip" of 4 minus the "zip" of x?

	Game One	Game Two	Game Three
Jerry	1st	4th	
Elaine			2nd
George			1st
Kramer	3rd	2nd	

21 The unfinished chart above records the results of 3 card games played by Jerry, Elaine, George, and Kramer. All four players played in each game, and there were no ties. If a player receives 5 points for first place, 3 points for second place, 1 point for third place, and no points for fourth place, what is the highest possible number of points that George could earn for all three games?

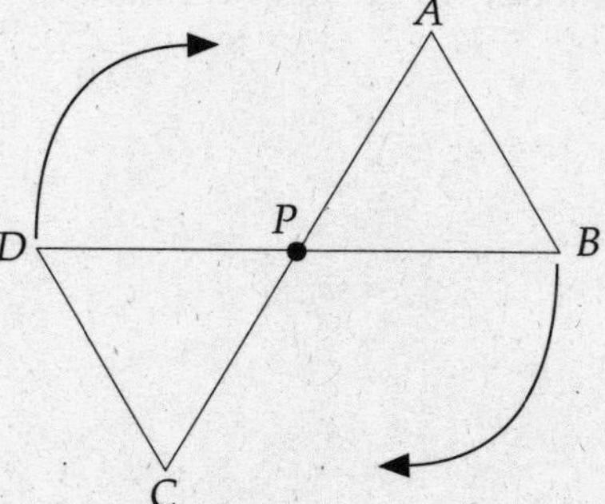

22 Two fan blades in the shape of equilateral triangles are rotated clockwise around Point P. How many degrees will the blades have rotated when Vertex D reaches the point where Vertex C is now?

HARD

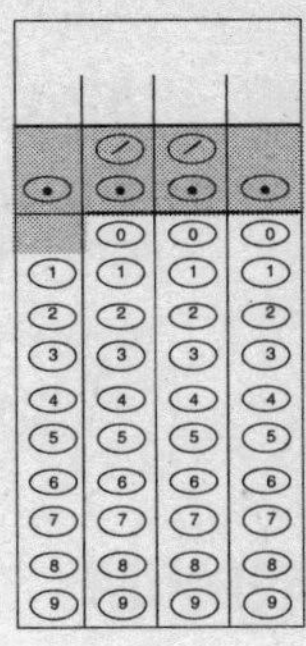

23 In a certain cereal, the ratio of wheat flakes to raisins to almonds is 14:2:1. If $\frac{1}{4}$ cup of cereal contains 5 raisins, how many more wheat flakes than almonds are there in 2 cups of cereal?

24 Seven neighbors who carpooled to work made a chart of the gallons of gas used by each member of the carpool during a certain time period. The number of gallons recorded on the chart were: 15, 10, 8, 8, 12, 14, and 17. The average number of gallons used was how much greater than the mode?

25 A cube with sides of 6 is spit into 8 smaller cubes of equal size. What is the total length of the edges of the 8 smaller cubes?

Answers and Explanations: Problem Set 20

EASY

16 8

The problem tells you that $y = 3$, so plug that into the equation and you get $2x - 9 = 7$. So $2x = 16$ and $x = 8$.

17 10

Write 170° next to a. There are 180 degrees in a line, so $b = 10$.

18 **3.50**

Kelly and Brandon spent \$28 per week for 2 weeks for a total of \$56. If they had rented the umbrella by the day, they would've spent 14 × \$4.25 for a total of \$59.50. That means they saved 59.50 – 56 = 3.50.

MEDIUM

19 **125**

$x = 3$, and $5^3 = 125$.

20 **255**

The "zip" of 4 = 4^4, or 256 (use your calculator). If we're looking for the greatest possible value, we should subtract the smallest thing we can—we have to use a positive integer (it says so in the question) so let's use the "zip" of 1: $1^1 = 1$. So the "zip" of 4 (256) minus the "zip" of 1(1) equals 255.

21 **13**

In Game One, George can't be 1st because Jerry is already 1st—so make George 2nd. In Game Two, make George 1st. Now count up George's points: Game One: 3 points. Game Two: 5 points. Game Three: 5 points. That's a total of 13 points.

22 **300**

Draw a circle around the fan blades, so that the vertices lie on the circle. Now imagine that you've stuck a pin in Point *P* and are rotating the blades around, clockwise. When *D* hits where *C* is now, it's gone almost all the way around, hasn't it? The rotation is 360° minus the number of degrees in angle *DPC*, and since the blades are equilateral triangles, that's 60°, so the rotation covers 300°.

This would have been a good question to guess on—if you could tell that the rotation was between 180 and 360, you might as well take a guess, since you won't lose any points for a wrong answer. Just guess quickly.

HARD

23 **260**

If there are 5 raisins in $\frac{1}{4}$ cup of cereal, there will be 40 raisins in 2 cups of cereal. ($\frac{1}{4} \times 8 = 2$ cups, and $5 \times 8 = 40$ raisins.) In the given ratio, the raisins were 2, and now there are 40. That means we need to multiply all the parts of the ratio by 20, since that's what we did to the raisins. That gives us a new ratio of 280: 40: 20, and all we have to do is subtract the almonds from the wheat flakes to get 260. Here's what your chart should look like:

	W	R	A
	14	2	1
$\frac{1}{4}$ c.		5	
2 c.	280	40	20

Sure, we could have figured out the wheat flakes and almonds in $\frac{1}{4}$ cup, but why bother—it would mean some ugly fractional messiness, and we might as well spend time dealing with 2 cups, since that's what the question wants anyway. Notice how a ratio can be multiplied by any number and still be the same ratio, as long as you multiply all the parts of the ratio by the same number, as we did here with 20.

24 **4**

Do the average on your calculator: $15 + 10 + 8 + 8 + 12 + 14 + 17 = 84$, and $84 \div 7 = 12$. The mode is the number that shows up the most, so the mode is 8. All that's left is $12 - 8 = 4$.

TIP: Average, median, and mode questions tend to be pretty easy, even when they're in the hard section. The only trick is knowing what a mode and a median is. So learn how to find them—it's not complicated—and you'll be able to pick up some relatively painless points.

25 **288**

It always helps to draw a diagram when they aren't nice enough to provide one. The original, uncut cube has sides of 6. If you divide it into 8 smaller cubes, the new sides will be 3. One cube has 12 edges, so the length of the edges of one cube is $3 \bullet 12 = 36$. Since you have 8 of these new cubes, multiply 36 by 8 and you get 288.

THE PRINCETON REVIEW NETWORK

The Princeton Review wants to provide you with the most up-to-date information you need whether you are preparing to take a test or apply to school. If you are using our books outside of the United States and have questions or comments, or simply want more information on our courses and the services The Princeton Review offers, please contact one of the following offices nearest to you.

- HONG KONG 852-517-3016
- JAPAN (Tokyo) 8133-463-1343
- KOREA (Seoul) 822-795-3028
- MEXICO CITY 011-525-358-0855
- PAKISTAN (Lahore) 92-42-872-315
- SAUDI ARABIA 413-548-6849 (a U.S. based number)
- SPAIN (Madrid) 341-446-5541
- TAIWAN (Taipei) 886-27511293

ABOUT THE AUTHOR

Cornelia Cocke has a BA from Dartmouth College and an MFA from Columbia University. She has taught SAT math to thousands of students since 1984.

Princeton Review

Guide to an Informed Mind

Has this ever happened to you?

You're at a party. An attractive person who you have been dying to meet comes over and says, "Man, does that woman have a Joan of Arc complex, or what?" and you, in a tone that is most suave, say "Uh," while your mind races wildly, "*Joan of Arc, OK, France, uh...damn,*" as the aforementioned attractive person smiles weakly and heads for the punch bowl.

No? How about this?

Your boss finally learns your name and says, "Ah, good to have you aboard. Do you like Renaissance painting?" You reply with an emphatic "Yes! I do!" to which she returns, "What's your favorite fresco?" You start stammering, glassy-eyed, your big moment passing you by as visions of soda pop dance through your brain.

CULTURESCOPE can help.

If you have gaps of knowledge big enough to drive a eighteen-wheeler through, The Princeton Review has the thing for you. It's called CULTURESCOPE. It's a book that can make people think you've read enough books to fill a semi, even if you haven't.

CULTURESCOPE covers everything: history, science, math, art, sports, geography, popular culture—it's all in there, and it's fun, because along with all of the great information there are quizzes, resource lists, fun statistics, wacky charts, and lots of pretty pictures.

It's coming soon to a bookstore near you.

You won't go away empty-headed.

To order this or any other Princeton Review title, call 1-800-793-BOOK.